AF864524

CIM Series in Mathematical Sciences

Series Editors

Irene Fonseca, Department of Mathematical Sciences, Carnegie Mellon University, Center for Nonlinear Analysis, Pittsburgh, PA, USA

Isabel Maria Narra Figueiredo, Dept of Math, University of Coimbra, COIMBRA, Portugal

The CIM Series in Mathematical Sciences is proudly published on behalf of and in collaboration with the International Center for Mathematics/Centro Internacional de Matemática (CIM). Proceedings, lecture course material from summer schools, and research monographs are welcome in the CIM Series.

Based in Portugal, this non-for-profit, privately-run association aims at developing and promoting research in mathematics. CIM is a member of ERCOM–European Research Centres on Mathematics and of IMSI–International Mathematics Sciences Institutes.

João Lopes Dias • Pedro Duarte •
José Pedro Gaivão • Silvius Klein • Telmo Peixe •
Jaqueline Siqueira • Maria Joana Torres
Editors

New Trends in Lyapunov Exponents

NTLE, Lisbon, Portugal, February 7–11, 2022

CIM

Springer

Editors
João Lopes Dias
Department of Mathematics,
CEMAPRE/REM/ISEG
University of Lisbon
Lisbon, Portugal

Pedro Duarte
Department of Mathematics and CMAFcIO,
Faculty of Sciences
University of Lisbon
Lisbon, Portugal

José Pedro Gaivão
Department of Mathematics,
CEMAPRE/REM/ISEG
University of Lisbon
Lisbon, Portugal

Silvius Klein
Department of Mathematics
Pontifical Catholic University of Rio de Janeiro
Rio de Janeiro, Brazil

Telmo Peixe
Department of Mathematics,
CEMAPRE/REM/ISEG
University of Lisbon
Lisbon, Portugal

Jaqueline Siqueira
Institute of Mathematics
Federal University of Rio de Janeiro
Rio de Janeiro, Brazil

Maria Joana Torres
CMAT and Department of Mathematics
University of Minho
Braga, Portugal

ISSN 2364-950X ISSN 2364-9518 (electronic)
CIM Series in Mathematical Sciences
ISBN 978-3-031-41315-5 ISBN 978-3-031-41316-2 (eBook)
https://doi.org/10.1007/978-3-031-41316-2

Mathematics Subject Classification: 37D25, 37D35, 37H15, 34F05

© The Editor(s) (if applicable) and The Author(s), under exclusive license to Springer Nature Switzerland AG 2023
This work is subject to copyright. All rights are solely and exclusively licensed by the Publisher, whether the whole or part of the material is concerned, specifically the rights of translation, reprinting, reuse of illustrations, recitation, broadcasting, reproduction on microfilms or in any other physical way, and transmission or information storage and retrieval, electronic adaptation, computer software, or by similar or dissimilar methodology now known or hereafter developed.
The use of general descriptive names, registered names, trademarks, service marks, etc. in this publication does not imply, even in the absence of a specific statement, that such names are exempt from the relevant protective laws and regulations and therefore free for general use.
The publisher, the authors, and the editors are safe to assume that the advice and information in this book are believed to be true and accurate at the date of publication. Neither the publisher nor the authors or the editors give a warranty, expressed or implied, with respect to the material contained herein or for any errors or omissions that may have been made. The publisher remains neutral with regard to jurisdictional claims in published maps and institutional affiliations.

This Springer imprint is published by the registered company Springer Nature Switzerland AG
The registered company address is: Gewerbestrasse 11, 6330 Cham, Switzerland

Paper in this product is recyclable.

Preface

The concept of characteristic exponent along an orbit of a dynamical system was introduced by Aleksandr Mikhailovich Lyapunov (1857–1918), who made fundamental contributions to the theory of differential equations and of dynamical systems. Known nowadays as Lyapunov exponent, this concept measures the sensitivity of an orbit to its initial condition. Roughly speaking, a negative Lyapunov exponent corresponds to stable orbit behavior, the kind characterized by Aleksandr Lyapunov, while positive Lyapunov exponents are associated with irregular or chaotic orbit behavior.

A systematic study of Lyapunov exponents followed the classical Multiplicative Ergodic Theorem of Valery Oseledets in 1965, where the concept is defined in the context of linear cocycles. A linear cocycle is a skew product dynamical system acting on a vector bundle, which preserves the linear bundle structure and induces a measure preserving dynamical system on the base. Lyapunov exponents quantify the average exponential growth of the iterates of the cocycle along fiber-invariant subspaces, which are called Oseledets subspaces.

An important class of examples of linear cocycles are the ones associated with discrete, one-dimensional ergodic Schrödinger operators. Such an operator is the discretized version of a quantum Hamiltonian. Its potential is given by a time series, that is, the potential is obtained by evaluating an observable along the orbit of an ergodic transformation.

The iterates of a linear cocycle can be thought of as a multiplicative (noncommutative) stochastic process. A relevant and difficult problem is to understand the statistical properties of such processes, under appropriate assumptions.

Some of the main topics of study in the theory of Lyapunov exponents are concerned with their positivity and simplicity; dichotomy between uniform hyperbolicity and zero Lyapunov exponents; regularity properties such as continuity, modulus of continuity or even smoothness; the structure of their level sets; their behavior on non-typical sets; generalizations of Oseledets' theorem to other settings and applications to fields such as Mathematical Physics, Differential Equations and Geometry.

This monograph contains a collection of survey articles describing recent research trends in these and related topics. The articles are authored by participants of the workshop “New trends in Lyapunov exponents” that took place between February 7 and 11, 2022, at ISEG-ULisboa (Lisbon School of Economics & Management of the Universidade de Lisboa) in Lisbon, Portugal. The workshop was part of the scientific activities organized within the research project PTDC/MAT-PUR/29126/2017 funded by FCT (Fundação para a Ciência e Tecnologia), Portugal. It also received partial support from ISEG and the following research centers: CMAT, CEMAPRE and CMAFcIO. For many participants, this gathering marked their first in-person international event after two years of travel restrictions resulting from the Covid-19 pandemic.

The editors are grateful to CIM (Centro Internacional de Matemática) for supporting the publication of this volume in the “CIM Series in Mathematical Sciences”.

Lisbon, Portugal — João Lopes Dias
Lisbon, Portugal — Pedro Duarte
Lisbon, Portugal — José Pedro Gaivão
Rio de Janeiro, Brazil — Silvius Klein
Lisbon, Portugal — Telmo Peixe
Rio de Janeiro, Brazil — Jaqueline Siqueira
Braga, Portugal — Maria Joana Torres

Contents

Lyapunov Exponents for Linear Homogeneous Differential Equations

Mário Bessa

Abstract We consider linear continuous-time cocycles $\Phi : \mathbb{R} \times M \to \mathrm{GL}(2, \mathbb{R})$ induced by second order linear homogeneous differential equations $\ddot{x}+\alpha(\varphi^t(\omega))\dot{x}+\beta(\varphi^t(\omega))x = 0$, where the coefficients α, β evolve along the orbit of a flow $\varphi^t : M \to M$ defined on a closed manifold M and $\omega \in M$. We are mainly interested in the Lyapunov exponents associated to most of the cocycles chosen when one allows variation of the parameters α and β. The topology used to compare perturbations turn to be crucial to the conclusions.

Keywords Differential equations · Linear cocycles · Linear differential systems · Multiplicative ergodic theorem · Lyapunov exponents

2010 Mathematics Subject Classification Primary: 34D08, 37H15; Secondary: 34A30, 37A20

1 Introduction

1.1 Linear Differential Systems

Let Φ^t_A be a matricial solution of the autonomous differential equation $\dot{U}(t) = A \cdot U(t)$ where A is a $n \times n$ matrix of the same order as $U(t)$. Given $v \in \mathbb{R}^n$, obtaining the asymptotic growth of the number

$$\frac{1}{t} \log \|\Phi^t_A \cdot v\| \tag{1}$$

is an exercise of finite dimensional spectral analysis. The Lyapunov spectrum is characterized by the Lyapunov exponents (the limit of (1) when $t \to \infty$) and

M. Bessa (✉)
Universidade Aberta - Departamento de Ciências e Tecnologia, Lisboa, Portugal
e-mail: mario.costa@uab.pt

© The Author(s), under exclusive license to Springer Nature Switzerland AG 2023

J. L. Dias et al. (eds.), *New Trends in Lyapunov Exponents*, CIM Series in Mathematical Sciences, https://doi.org/10.1007/978-3-031-41316-2_1

its eigendirections. When a perturbation B of A is allowed, say by considering B uniformly near A, the perturbed system $\dot{U}(t) = B \cdot U(t)$ originates the solution Φ_B^t exhibiting a well-understood behaviour [36]. The problem starts to gain enormous complexity when we consider the non-autonomous case $\dot{U}(t) = A(t) \cdot U(t)$, where $A(t)$ depends on the parameter t. The calculation of Lyapunov exponents, as well as their stability, turns out to be a substantially more difficult issue. A classical way of looking at non-autonomous linear differential equations is to consider linear differential systems i.e. linear continuous-time cocycles.

1.2 Kinetic Cocycles

It has been known by *Liouville theory* (see e.g. [45]), established almost two centuries ago, that there are huge constraints when we try to apply analytical methods to integrate most functions. As we cannot always have explicit solutions a qualitative approach to understand the asymptotic behaviour of solutions of differential equations proved to be an efficient approach to deal with this difficulty. We intend to analyse the asymptotic behaviour of solutions of second order linear homogeneous differential equations of the form

$$\ddot{x}(t) + \alpha(\varphi^t(\omega))\dot{x}(t) + \beta(\varphi^t(\omega))x(t) = 0, \qquad (2)$$

with coefficients α and β displaying a certain regularity (L^p or C^r with $r \geq 0$), varying along the orbits of a flow φ^t and admitting a small perturbation on the parameters α and β. This flow φ^t is usually consider to be aperiodic and so for each orbit we obtain a particular differential equation which results in dealing with infinitely many differential equations at the same time. We will try to describe the Lyapunov spectrum of a linear cocycle associated with (2) when some perturbation on its coefficients is made. The details about these type of cocycles will be presented later on Sect. 2.2 but the idea is very simple. We consider a flow $\varphi^t \colon M \to M$ preserving a measure defined in M and a linear variational equation $\dot{U}(\omega, t) = A(\varphi^t(\omega)) \cdot U(\omega, t)$ with generator

$$\begin{aligned} A \colon M &\longrightarrow \mathbb{R}^{2\times 2} \\ \omega &\longmapsto \begin{pmatrix} 0 & 1 \\ -\beta(\omega) & -\alpha(\omega) \end{pmatrix}. \end{aligned} \qquad (3)$$

Hence, the infinitesimal generator A is of a particular type after all. The flow φ^t will label a certain differential equation where A captures its coefficients.

1.3 The Harmonic Oscillator

Equation (2) represents the simple damped harmonic oscillator free from external forces where α (frictional force) and β (frequency of the oscillator) are functions depending on $\omega \in M$ described by the flow $\varphi^t : M \to M$ for $t \in \mathbb{R}$. When the frictional force and the frequency of the oscillator are constant or, more generally, when α and β are first integrals with respect to φ^t, then (2) can be easily solved. When this is not the case, explicit solutions could be difficult to obtain. When $\alpha \neq 0$ (damped case) we have the solution $\Phi^t_A \in \mathrm{GL}(2, \mathbb{R})$ and when $\alpha = 0$ (the frictionless case) we have $\Phi^t_A \in \mathrm{SL}(2, \mathbb{R})$. Clearly, a perturbation theory for the frictionless case deserves some kind of care because perturbations will have to maintain $\alpha = 0$.

1.4 The Main Goal

Given (3) and fixing the *position* and *velocity* $(x(0), \dot{x}(0))$ we are interested in describing the Lyapunov spectrum of Φ^t_A when $t \to \infty$ of the pair $(x(t), \dot{x}(t))$. More particularly, we intend to address the following problem:

- Fixing a certain regularity of the parameters α and β (L^p, L^∞, C^0, C^1, ...) and
- providing the parameter space with a conforming topology we ask:
- For the ‘majority’ of parameters considered (dense/residual/open+dense) what kind of Lyapunov spectrum do we expect to have?

In Sect. 3 we discuss the case when parameters evolve on L^p, in Sect. 4 we consider parameters on C^0 and finally, in Sect. 5 we consider the parameters evolving on C^r. The only case where there is already results in literature is the L^p one. Hence, considering the C^0 and the C^r cases we will only address some open questions. Finally, in Sect. 6 we consider a particular model of a third order linear homogeneous differential equation and follow [18] to show how to remove zero Lyapunov exponents on a partial hyperbolic cocycle by a small C^0 perturbation.

2 Kinetic Linear Cocycles

2.1 Linear Cocycles

Let $(M, \mathcal{M}, \mu)$ be a probability space and let $\varphi \colon \mathbb{R} \times M \to M$ be a *measurable flow* in the sense that it is a measurable map and

(1) $\varphi^t : M \to M$ given by $\varphi^t(\omega) = \varphi(t, \omega)$ preserves the measure μ for all $t \in \mathbb{R}$;
(2) $\varphi^0 = \mathrm{Id}_M$ and $\varphi^{t+s} = \varphi^t \circ \varphi^s$ for all $t, s \in \mathbb{R}$.

Unless stated otherwise we will consider that the flow is ergodic in the usual sense that there exist no invariant sets except zero measure sets and their complements. Let $\mathcal{B}(X)$ be the Borel σ-algebra of a topological space X. A continuous-time linear *random dynamical system* on $(\mathbb{R}^2, \mathcal{B}(\mathbb{R}^2))$, or a continuous-time *linear cocycle*, over φ is a $(\mathcal{B}(\mathbb{R}) \otimes \mathcal{M}/\mathcal{B}(\mathrm{GL}(2, \mathbb{R}))$-measurable map

$$\Phi : \mathbb{R} \times M \to \mathrm{GL}(2, \mathbb{R})$$

such that the mappings $\Phi(t, \omega)$ form a cocycle over φ, that is:

(1) $\Phi^0(\omega) = \mathrm{Id}$ for all $\omega \in M$;
(2) $\Phi^{t+s}(\omega) = \Phi^t(\varphi^s(\omega)) \circ \Phi^s(\omega)$, for all $s, t \in \mathbb{R}$ and $\omega \in M$,

and $t \mapsto \Phi^t(\omega)$ is continuous for all $\omega \in M$. We recall that having $\omega \mapsto \Phi^t(\omega)$ measurable for each $t \in \mathbb{R}$ and $t \mapsto \Phi^t(\omega)$ continuous for all $\omega \in M$ implies that Φ is measurable in the product measure space. We also call these objects *linear differential systems*.

2.2 *Kinetic Linear Cocycles*

As we already said, in Sect. 1.3, the cocycles we consider are motivated by the non-autonomous linear homogeneous differential equation which describes a motion of the damped harmonic oscillator as the 'simple pendulum' along the path $(\varphi^t(\omega))_{t\in\mathbb{R}}$, with $\omega \in M$ described by the flow φ. Let $K \subset \mathbb{R}^{2\times 2}$ be the set of 2×2 matrices written in the form of

$$A = \begin{pmatrix} 0 & 1 \\ b & a \end{pmatrix} \tag{4}$$

for real numbers a, b, and denote by $\mathcal{G}$ the set of measurable applications $A : M \to \mathbb{R}^{2\times 2}$. Denote also by $\mathcal{K} \subset \mathcal{G}$ the set of *kinetic* measurable applications $A : M \to K$. We also identify two applications on $\mathcal{G}$ that coincide on μ-a.e. in M. Take measurable maps $\alpha : M \to \mathbb{R}$ and $\beta : M \to \mathbb{R}$. Consider the differential equation given in (2). Let $y(t) = \dot{x}(t)$ and rewrite (2) as the following vectorial linear system

$$\dot{X} = A(\varphi^t(\omega)) \cdot X, \tag{5}$$

where $X = X(t) = (x(t), y(t))^T = (x(t), \dot{x}(t))^T$ and $A \in \mathcal{K}$ is given by (3). It follows from [5, Thm. 2.2.2] (see also Lemma 2.2.5 and Example 2.2.8 in this reference) that if $A \in \mathcal{G}^1 =: \mathcal{G} \cap L^1(\mu)$, i.e. $\int_M \|A\| \, d\mu < \infty$, then it generates a unique linear differential system Φ_A satisfying

$$\Phi^t_A(\omega) = \mathrm{Id} + \int_0^t A(\varphi^s(\omega)) \cdot \Phi^s_A(\omega)\, ds. \tag{6}$$

The solution $\Phi^t_A(\omega)$ defined in (6) is called *mild solution* or *Carathéodory solution*. Given an initial condition $X(0) = v \in \mathbb{R}^2$, we say that $t \mapsto \Phi^t_A(\omega)v$ solves or is a solution of (5), or that (5) generates $\Phi^t_A(\omega)$. Note that $\Phi^0_A(\omega)v = v$ for all $\omega \in M$ and $v \in \mathbb{R}^2$. If the solution (6) is differentiable in time (i.e. with respect to t) and satisfies for all t

$$\frac{d}{dt}\Phi^t_A(\omega)v = A(\varphi^t(\omega)) \cdot \Phi^t_A(\omega)v \quad \text{and} \quad \Phi^0_A(\omega)v = v, \tag{7}$$

then it is called a *classical solution* of (5). Classical solutions arise when we consider $A : M \to K$ continuous. Of course that $t \mapsto \Phi^t_A(\omega)v$ is continuous for all ω and v. Due to (7) we call $A : M \to K$ a *kinetic infinitesimal generator* of Φ_A. Sometimes, due to the relation between A and Φ_A, we refer to both A and Φ_A as a kinetic linear cocycle or kinetic linear differential system. If (5) has initial condition $X(0) = v$ then $\Phi^0_A(\omega)v = v$ and $X(t) = \Phi^t_A(\omega)v$. Let $\mathcal{K}_0 \subset \mathcal{K}$ stand for the *traceless kinetic cocycles* induced from matrices as in (4) imposing the constraint $a = 0$. Let $\mathcal{K}^1 = \mathcal{K} \cap L^1(\mu) \subset \mathcal{G}^1$ and let $\mathcal{K}^1_0 = \mathcal{K}_0 \cap L^1(\mu) \subset \mathcal{K}^1$.

2.3 Topologies

Now we will define the topologies we are going to consider in the sequel.

2.3.1 The L^p Topology

We now define an L^p-like topology generated by a metric that compares the infinitesimal generators on $\mathcal{G}$. For $1 \leq p < \infty$ and $A, B \in \mathcal{G}$ we set

$$\hat{\sigma}_p(A, B) := \begin{cases} \left(\int_M \|A(\omega) - B(\omega)\|^p \, d\mu(\omega) \right)^{\frac{1}{p}}, \\ \infty \ \text{ if the above integral does not exists,} \end{cases}$$

and define

$$\sigma_p(A, B) := \begin{cases} \frac{\hat{\sigma}_p(A,B)}{1+\hat{\sigma}_p(A,B)}, & \text{if } \hat{\sigma}_p(A, B) < \infty \\ 1, & \text{if } \hat{\sigma}_p(A, B) = \infty \end{cases}.$$

Clearly, σ_p is a distance in $\mathcal{G}$. The following topological results were proved in [2].

Proposition 2.1 *Consider* $1 \leq p < \infty$. *Then:*

(i) $\sigma_p(A, B) \leq \sigma_q(A, B)$ *for all* $q \geq p$ *and all* $A, B \in \mathcal{G}$.
(ii) *If* $A \in \mathcal{G}^1$ *then for any* $B \in \mathcal{G}$ *satisfying* $\sigma_p(A, B) < 1$ *we have* $B \in \mathcal{G}^1$. *Therefore,* $\sup_{0 \leq t \leq 1} \log^+ \|\Phi^t_B(\omega)^{\pm 1}\| \in L^1(\mu)$.
(iii) *The sets* $(\mathcal{K}^1, \sigma_p)$ *and* $(\mathcal{K}^1_0, \sigma_p)$ *are closed, for all* $1 \leq p < \infty$.
(iv) *For all* $1 \leq p < \infty$, $(\mathcal{K}^1, \sigma_p)$ *and* $(\mathcal{K}^1_0, \sigma_p)$ *are complete metric spaces and, therefore Baire spaces.*

Next result is elementary in measure theory and captures the crucial idea which allows making huge perturbations on the uniform norm but small perturbations in the L^p-norm as long the support is small in measure. For the proof see [3].

Lemma 2.2 *Given* $A \in \mathcal{G}^1$ *and* $\epsilon > 0$ *there exists* $\delta > 0$ *such that if* $\mathcal{F} \in \mathcal{M}$ *and* $\mu(\mathcal{F}) < \delta$, *then* $\int_{\mathcal{F}} \|A(\omega)\|\, d\mu(\omega) < \epsilon$.

2.3.2 Uniform Topologies

Now we consider that the kinetic infinitesimal generators $A : M \to \mathbb{R}^{2\times 2}$ are in L^∞ or are in C^0. The first is denoted by $L^\infty(M, \mathcal{K})$ and the second by $C^0(M, \mathcal{K})$. Clearly, $C^0(M, \mathcal{K}) \subset L^\infty(M, \mathcal{K}) \subset \mathcal{K}^1$. We also consider traceless infinitesimal generators $C^0_0(M, \mathcal{K}) \subset L^\infty_0(M, \mathcal{K}) \subset \mathcal{K}^1_0$.

We endow $L^\infty(M, \mathbb{R}^{2\times 2})$ with the L^∞ metric defined by

$$\|A - B\|_\infty = \operatorname*{ess\,sup}_{\omega \in M} \|A(\omega) - B(\omega)\|$$

where $A, B \in L^\infty(M, \mathbb{R}^{2\times 2})$. We also endow $C^0(M, \mathbb{R}^{2\times 2})$ with the C^0 metric defined by

$$\|A - B\|_0 = \max_{\omega \in M} \|A(\omega) - B(\omega)\|$$

where $A, B \in C^0(M, \mathbb{R}^{2\times 2})$. We also make use of the uniform operators norm to compare solutions given a fixed $t > 0$ like

$$\|\Phi^t_A - \Phi^t_B\|_0 = \max_{\omega \in M} \|\Phi^t_A(\omega) - \Phi^t_B(\omega)\| \quad \text{or} \quad \|\Phi^t_A - \Phi^t_B\|_\infty = \operatorname*{ess\,sup}_{\omega \in M} \|\Phi^t_A(\omega) - \Phi^t_B(\omega)\|.$$

Both $(C^0(M, \mathbb{R}^{2\times 2}), \|\cdot\|_0)$ and $(L^\infty(M, \mathbb{R}^{2\times 2}), \|\cdot\|_\infty)$ are complete metric spaces and, therefore, Baire spaces. The set $L^\infty(M, \mathcal{K})$ is L^∞-closed and the set $C^0(M, \mathcal{K})$ is C^0-closed. Moreover, the set $L^\infty(M, \mathcal{K}_0)$ is L^∞-closed and the set $C^0(M, \mathcal{K}_0)$ is C^0-closed.

2.3.3 The C^r Topology

Finally, we consider that the kinetic infinitesimal generators $A : M \to \mathbb{R}^{2\times 2}$ are $C^{r,\nu}$ (i.e. are in $C^{r+\nu}$) a set we denote by $C^{r,\nu}(M, \mathcal{K})$, where $r \in \mathbb{N} \cup \{0\}$ and $\nu \in [0, 1]$. We endow $C^{r,\nu}(M, \mathbb{R}^{2\times 2})$ with the $C^{r,\nu}$-topology defined using the norm

$$\|A\|_{r,\nu} = \sup_{0\le j\le r} \sup_{x\in M} \|D^j A(x)\| + \sup_{x\neq y} \frac{\|A(x) - A(y)\|}{\|x - y\|^\nu}, \tag{8}$$

where $A \in C^{r,\nu}(M, \mathbb{R}^{2\times 2})$ and $x, y \in M$. Let us also mention that it is enough to consider the case when $\nu = 1$ (i.e. A is Lipschitz). In fact, if A is ν-Hölder continuous with respect to the metric $d(\cdot, \cdot)$ then it is Lipschitz with respect to the metric $d(\cdot, \cdot)^\nu$. Hence, up to a change of metric we may assume that A is Lipschitz and we will do so throughout the presentation.

2.4 *Lyapunov Exponents*

Notice that if $A \in \mathcal{K}^1$ then the cocycle Φ_A satisfies the following *integrability condition*

$$\sup_{0\le t\le 1} \log^+ \|\Phi_A^t(\omega)^{\pm 1}\| \in L^1(\mu), \tag{9}$$

where $\log^+ = \max\{0, \log\}$. In fact, take ω in the full measure φ^t-invariant subset of M where $t \mapsto A(\varphi^t(\omega))$ is locally integrable. By (6) and by Grönwall's inequality (see [5]) we get

$$\sup_{0\le t\le T} \log^+ \|\Phi_A(t, \omega)^{\pm 1}\| \le \int_0^T \|A(\varphi^s(\omega))\|\, ds =: \psi(\omega, T). \tag{10}$$

By Arnold [5, Lemma 2.2.5] we have $\psi(\cdot, T) \in L^1(\mu)$, hence (9) holds. Fubini's theorem allow us also to obtain from (10) that:

$$\begin{aligned}\int_M \log^+ \|\Phi_A(t, \omega)^{\pm 1}\|\, d\mu(\omega) &\le \int_M \int_0^t \|A(\varphi^s(\omega))\|\, ds\, d\mu(\omega)\\ &= \int_0^t \int_M \|A(\varphi^s(\omega))\|\, d\mu(\omega)\, ds = t\|A\|_1.\end{aligned}$$

If $A \in \mathcal{G}^1$ then the integrability condition (9) holds and Oseledets theorem (see e.g. [5, 44]) gives that for μ-a.e. $\omega \in M$, there exists a Φ_A-invariant splitting called

Oseledets splitting of the fiber $\mathbb{R}^2_\omega = E^1_\omega \oplus E^2_\omega$ and real numbers called *Lyapunov exponents* $\lambda_1(A, \omega) \geq \lambda_2(A, \omega)$, such that:

$$\lambda_i(A, \omega) = \lim_{t \to \pm\infty} \frac{1}{t} \log \|\Phi^t_A(\omega) v^i\|, \tag{11}$$

for any $v^i \in E^i_\omega \setminus \{\vec{0}\}$ and $i = 1, 2$. Furthermore, given subspaces E^1_ω and E^2_ω, the angle between them along the orbit has subexponential growth, meaning that

$$\lim_{t \to \pm\infty} \frac{1}{t} \log \sin \measuredangle(E^1_{\varphi^t(\omega)}, E^2_{\varphi^t(\omega)}) = 0. \tag{12}$$

If the flow φ^t is ergodic, then the real numbers (11) and the dimensions of the associated subbundles are constant μ almost everywhere and we will denote them by $\lambda_1(A)$ and $\lambda_2(A)$. We say that A has *trivial Lyapunov spectrum* or *one-point Lyapunov spectrum* (respectively *simple Lyapunov spectrum*) if for μ a.e. $\omega \in M$, $\lambda_1(A, \omega) = \lambda_2(A, \omega)$ (respectively $\lambda_1(A, \omega) > \lambda_2(A, \omega)$).

2.5 *The Search for Positive Lyapunov Exponents*

A positive Lyapunov exponent gives us the average exponential rate of divergence of two neighboring orbits whereas a negative Lyapunov exponent gives us the average exponential rate of convergence of two neighboring orbits. Zero Lyapunov exponents gives us the lack of any kind of asymptotic exponential behaviour. The nonuniform hyperbolic theory [13] guarantees a invariant manifold theory in the presence of non-zero Lyapunov exponents. These geometric considerations are the basis of most of the central results in today's dynamical systems. Hence, there can be no doubt that pursuing non-zero Lyapunov exponents is an important feature in dynamics over the last 60 years (see e.g. [40]). Some criteria for the positivity of the Lyapunov exponents were obtained by Cornelis and Wojtkowski [26], and Ledrappier [35] and later Knill [42] and Nerurkar [41] showed that for a C^0-dense set of certain cocycles we have non-zero Lyapunov exponents. Arnold and Cong [8] proved the L^p-denseness of positive Lyapunov exponents and their technique was generalized in [20]. The use of a classical method developed by Moser and linked to the concept of rotation number allowed Fabbri and Johnson to obtain abundance of positive Lyapunov exponents for linear differential systems evolving on $\mathrm{SL}(2, \mathbb{R})$ on the fiber and displaying a translation on the torus on the base (see [29, 31, 32] and also the work with Zampogni [33]). Due to area-preserving invariance, obtaining a positive Lyapunov exponent $\lambda > 0$ in $\mathrm{SL}(2, \mathbb{R})$ allows us to obtain a negative Lyapunov exponent $-\lambda < 0$ and thus all Lyapunov exponents are different. A variety of results guaranteeing the positivity of Lyapunov exponents for strong topologies established recently bring out different new approaches [21, 24, 28, 47, 48]. As an example, in [11], Avila obtained prevalence of simple spectrum in a rather wide range of topologies and on the two dimensional case. The topology used to compare perturbations turn to be crucial to the conclusions.

3 The L^p Case

3.1 Towards Zero Lyapunov Exponents

The L^p-generic description of the Lyapunov spectrum for general linear differential systems was first studied in [20] by the author and Vilarinho following the pioneering approaches by Arnold-Cong and Arbieto-Bochi [4, 7]. In [20] was proved that the class of *accessible* (*twisting*) linear differential systems, a wider class that includes cocycles that evolve in $\mathrm{GL}(n, \mathbb{R})$, $\mathrm{SL}(n, \mathbb{R})$ and $\mathrm{Sp}(2n, \mathbb{R})$, have a trivial Lyapunov spectrum L^p-generically. If we consider the stronger C^0-norm, then Millionshchikov's work [40] in the late sixties shows that the generic behaviour changes. We will consider this C^0 case in Sect. 4. For the time being we now describe the recent results by the author, Amaro and Vilarinho. In rough terms in [2] was obtained that for an L^p-generic choice of a kinetic linear differential system (as in (2)) and for almost every driving realization, no matter what position and momentum $(x(0), \dot{x}(0))$ we chose as initial conditions, the asymptotic exponential behaviour of the solutions will be the same.

In [2] was proved the following result which is the kinetic version of [20, Theorem 1].

Theorem 1 ([2]) *For all* $1 \leq p < \infty$ *there exists a* σ_p*-residual subset* $\mathcal{R} \in \mathcal{K}^1$ *such that any* $A \in \mathcal{R}$ *has one-point spectrum.*

The two main components to prove Theorem 1 are Propositions 3.1 and 3.2. Once we establish these two results the proof of Theorem 1 is easily obtained. To prove Proposition 3.1 we used Lemma 2.2 and through a perturbation we caused a rotational effect of Oseledets directions. Unfortunately, rotating in $\mathcal{K}^1$ is much more difficult and the way to overcome this problem is to induce rotations via translations in the projective plane. In summary we perform (fake) rotations but remain in the kinetic class. Once we know how to 'rotate', a classical Mañé argument (see e.g. [14, 15, 22, 23]) allows us to get:

Proposition 3.1 *Given* $A \in \mathcal{K}^1$ *and* $\varepsilon, \delta > 0$*, there exists* $B \in \mathcal{K}^1$ *such that* $\sigma_1(A, B) < \varepsilon$ *and*

$$\lambda_1(B) \leq \frac{\lambda_1(A) + \lambda_2(A)}{2} + \delta. \tag{13}$$

Inequality (13) is used to decrease the upper Lyapunov exponent of a perturbation of the original linear differential system.

Finally, in Proposition 3.2 we obtain the upper semi-continuity of the top Lyapunov exponent function with respect to the L^p topology. We define the function

$$\begin{aligned} \mathscr{L}\colon (\mathcal{G}^1, \sigma_p) &\longrightarrow \mathbb{R} \\ A &\longmapsto \int_M \lambda_1(A, \omega)\, d\mu(\omega). \end{aligned} \tag{14}$$

Clearly, if μ is ergodic for the flow φ^t we have $\mathscr{L}(A) = \lambda_1(A)$.

Proposition 3.2 *For all $1 \leq p < \infty$, the function $\mathscr{L}$ is upper semicontinuous when we endow $\mathcal{G}^1$ with the σ_p-topology, that is, for all $A \in \mathcal{G}^1$ and $\varepsilon > 0$ there is $\delta > 0$ such that $\sigma_p(A, B) < \delta$ implies $\mathscr{L}(B) < \mathscr{L}(A) + \varepsilon$.*

In order to prove that $\mathscr{L}$ is upper semi-continuous when $\mathcal{G}^1$ is endowed with the σ_p metric defined in Sect. 2.3.1 we must deal with the two main continuity-like problems:

Step 1 The first had already appeared [4, 20]. Indeed, it was the main step in [4] in order to improve from L^p-dense (cf. [7]) to L^p-residual. We are talking about the way it is used a simple measure-theoretical result (in brief terms that L^1 functions are L^∞ in a large part of the domain) to still guarantee continuity properties even under L^p-regularity.

Step 2 The second one is also a problem on continuity but a bit more difficult. This time on continuous dependence of solutions of differential equations. Notice that the function $\mathscr{L}$ in (14) is defined using the Lyapunov exponent which in turn is defined using the solution Φ^t_A and not the infinitesimal generator which is precisely the input on the σ_p-topology. So we need to get that solutions Φ^t_A and Φ^t_B are σ_p-near if its corresponding infinitesimal generators A and B are σ_p-near.

3.2 Towards Non-zero Lyapunov Exponents

The L^p-dense characterization of the Lyapunov spectrum for general linear differential systems was also considered in [20] generalizing this time the work by Arnold-Cong [8]. In [20] was proved that the class of *accessible* (a *twisting* type of property) and *saddle-conservative* (a *pinching* type of property) linear differential systems, a wider class that includes again cocycles that evolve in $\mathrm{GL}(n, \mathbb{R})$, $\mathrm{SL}(n, \mathbb{R})$ and $\mathrm{Sp}(2n, \mathbb{R})$, have simple Lyapunov spectrum L^p-densely. Recently, in [3], was proved a corresponding result for kinetic linear differential systems. Hence, we get in particular that the residual of Theorem 1 cannot contain L^p-open sets. We state now this result which establishes the existence of a σ_p-dense subset of $\mathcal{K}^1$ displaying simple spectrum:

Theorem 2 *Let $\varphi^t : M \to M$ be ergodic. For any $A \in \mathcal{K}^1$, $1 \leq p < \infty$ and $\epsilon > 0$, there exists $B \in \mathcal{K}^1$ exhibiting simple Lyapunov spectrum satisfying $\sigma_p(A, B) < \epsilon$.*

The ideas to prove Theorem 2 borrow arguments present in [8, 20] where were obtained similar results for more general cocycles (some input of works like [12, 19] are also present). Nevertheless, and once again, the constraint of perturbing in the kinetic context ultimately makes our task difficult. There are two perturbations to achieve non-zero Lyapunov exponents: one was in a certain sense already used to proof Theorem 1 (rotation), the other (stretch) is a new ingredient here. We emphasize that performing these perturbations (rotation and stretch) is thorough because we are not working in the broad family of cocycles that satisfy the *accessibility* and *saddle-conservative* which allow these processes to be carried out in a less demanding manner (see [4, 8, 20, 22, 23]). In [3] was used a slightly different approach when compared to previous works [6, 8, 20, 27]. To avoid overlapping in the perturbations, the base flow was encoded through a special flow in a Kakutani Castle (as in [1, 46]). We now describe the main steps done in [3] towards the proof of Theorem 2:

Step 1 We begin by considering two time-1 concatenated thin flowboxes $\mathcal{V}_R$ and $\mathcal{V}_S$ and so $\mathcal{V}_R \cup \mathcal{V}_S$ will be a time-2 flowbox: in $\mathcal{V}_R$ a rotational perturbation was made such that for each point ω entering in $\mathcal{V}_R$ we rotate in time-1 a vector v_ω into a fixed special vector v. The vector v_ω codifies the Oseledets direction associated to a cocycle A with trivial spectrum and v will be used to create another Oseledets direction.

Step 2 Then, in $\mathcal{V}_S$, we stretch the vector v. This small stretch allows to obtain simple Lyapunov spectrum but some control on the trace is needed in order to control the sum of the two Lyapunov exponents.

Step 3 Finally, ergodicity is used to compute the Lyapunov exponents of points who will inevitably have to return to $\mathcal{V}_R \cup \mathcal{V}_S$ infinitely many times.

Since all the perturbation we made are traceless, we get from Theorem 2 the conservative version:

Corollary 1 *Let $\varphi^t : M \to M$ be ergodic. For any $A \in \mathcal{K}_0^1$, $1 \leq p < \infty$ and $\epsilon > 0$, there exists $B \in \mathcal{K}_0^1$ exhibiting non-zero Lyapunov exponents satisfying $\sigma_p(A, B) < \epsilon$.*

4 The C^0 Case

We continue the main goal of understanding the asymptotic behaviour of the solutions of (2) but now allowing C^0-small perturbations on the parameters taking into account the topology considered in Sect. 2.3.2. There is significant literature on this issue [14, 16, 17, 29–33, 40, 43] addressing a more general or less general scenario than the one we are considering here.

4.1 Hyperbolicity

We say that the splitting $E_M = E^1 \oplus E^2$ is Φ^t_A-invariant if $\Phi^t_A(\omega)E^i_\omega = E^i_{\varphi^t(\omega)}$ for $i = 1, 2$ and $t \in \mathbb{R}$. We say that $E^1 \oplus E^2$ is a (C, σ)-dominated splitting for Φ^t_A if it is Φ^t_A-invariant and there exist $C > 0$ and $\sigma \in]0, 1[$ such that, for all $\omega \in M$ and $t \geq 0$ we have:

$$\frac{\|\Phi^t_A(\omega)|_{E^2_\omega}\|}{\mathfrak{m}(\Phi^t_A(\omega)|_{E^1_\omega})} \leq C\,\sigma^t,$$

where $\mathfrak{m}(\cdot)$ denotes the co-norm of an operator, that is $\mathfrak{m}(A) = \|A^{-1}\|^{-1}$. We say that the subbundle E is *hyperbolic* if either $\|\Phi^{-t}_A(\omega)\cdot u\| \leq C\,\sigma^t$ (expanding), for all $\omega \in M$, $t \geq 0$ and any unit vector $u \in E_\omega$, or $\|\Phi^t_A(\omega)\cdot u\| \leq C\,\sigma^t$ (contracting), for all $\omega \in M$, $t \geq 0$ and any unit vector $u \in E_\omega$.

4.2 A Mechanism for Obtaining Zero Lyapunov Exponents

Two results deserve particular attention from us since they are quite related to the problem in study. Firstly, motivated by the Mañé-Bochi dichotomy (see [22, 38, 39]) the author proved in [14] that C^0-generically traceless 2-dimensional linear differential systems over a conservative flow have, for almost every point, a dominated splitting (which, in fact is hyperbolic) or alse has a trivial Lyapunov spectrum. Secondly, in [29] Fabbri proved that Schrödinger cocycles[1] with a quasi-periodic potential and over a certain flow on the torus display a similar dichotomy.

Both results are somehow related to our kinetic frictionless case setting but in [29] the linear differential systems evolve in a much more rigid family and in [14] the linear differential system evolve in a broader family. What we intend is to address the mechanism for obtaining zero Lyapunov exponents developed in the Mañé-Bochi dichotomy. A complete treatment on the C^0-case was done in [14, 16] after the discrete approach done in [22, 23]. This mechanism is supported in perturbations which are rotations and, as we already saw, 'rotating' is a much more delicate issue in the kinetic setting. We make an interlude to focus on something that might already have been explained in the previous section: whenever our arguments require accessibility in projective space (rotations are precisely a way of accessing from one direction to another) having low dimension plays a crucial role. It is really quite rare to be able to carry out these kinds of arguments in dimension > 2 precisely because of a lack of enough degrees of freedom (see Sect. 6).

[1] In (2) take $\alpha = 0$ and $\beta = -E + Q(\varphi^t(\omega))$ where E is the energy and Q a quasi-periodic potential.

Infinitesimal generators like A in (3) generate a particular class of solutions: when $\alpha \neq 0$ the solutions of (7) evolve on a subclass of $\mathrm{GL}(2, \mathbb{R})$ and when $\alpha = 0$ the solutions evolve on a subclass of $\mathrm{SL}(2, \mathbb{R})$. Neither of these two subclasses are subgroups of $\mathrm{GL}(2, \mathbb{R})$. Hence, a careful study should be made considering the constraint that perturbations must belong to our kinetic class and not to the broader class of cocycles evolving in $\mathrm{GL}(2, \mathbb{R})$ or even in $\mathrm{SL}(2, \mathbb{R})$ (see e.g. [9, 10, 16, 34, 37]).

Our context is the following:

Base dynamics: We will consider a flow φ^t on a Hausdorff topological space M leaving invariant a Borel regular probability measure μ.

Fiber dynamics: We consider the infinitesimal generator of the form (3) where α and β are C^0 functions.

Topology: We endow $A\colon M \to \mathcal{K}$ with the C^0 norm described in Sect. 2.3.2.

The question of knowing the C^0-generic asymptotic behaviour of linear differential systems arising from equations like (2) is a work in progress. Nevertheless, we may ask the following two questions:

Open Question 1 *Is the following dichotomy:*

- *there exists a single Lyapunov exponent for* $\Phi_A^t(\omega)$ *or else*
- *the splitting along the orbit of* ω *is dominated,*

true for μ*-a.e.* ω *and a* C^0*-generic kinetic* C^0 *linear differential system* A*?*

Then we may also ask about the frictionless case:

Open Question 2 *Is the following dichotomy:*

- *all Lyapunov exponents of* $\Phi_A^t(\omega)$ *vanishes or else*
- *the splitting along the orbit of* ω *is hyperbolic,*

true for μ*-a.e.* ω *and a* C^0*-generic traceless kinetic* C^0 *linear differential system* A*?*

5 The C^r $(r > 0)$ Case

As we saw in Sect. 3.2, using a weak topology we get a dense set of parameters α, β such that the Lyapunov spectrum is simple. Yet, by Sect. 3.1 this dense set cannot be open. Then, in Sect. 4 we saw that outside the global hyperbolic setting there is always a set, relevant from the point of view of the measure, to which we have trivial Lyapunov spectrum. Therefore, simplicity of the spectrum is more plausible to exist stably when we pick a stronger topology. When considering a C^r $(r > 0)$ topology there are several results in the literature that point in the direction of robustness of simple spectrum [21, 24, 25, 47].

We now consider that Eqs. (2) in Sect. 2.2 are modeled by a linear cocycle Φ_A^t associated to $A \in C^{r,\nu}(M, \mathcal{K})$. Specifically, our context is as follows:

Base dynamics: Let M be a compact Riemannian manifold, φ^t a $C^{1+\alpha}$-flow on M ($\alpha > 0$) preserving a probability measure μ which is also hyperbolic and has local product structure (see [47]).

Fiber dynamics: We consider the infinitesimal generator of the form (3) where α and β are $C^{r,\nu}$ functions.

Topology: We endow $A\colon M \to \mathcal{K}$ with the $C^{r,\nu}$ norm (8) described in Sect. 2.3.3.

One of the main steps in Viana's proof [47] is an elegant lemma asserting that the holonomy maps are submersions as function of $A \in C^{r,\nu}(M, G)$, where G is a subgroup of $\mathrm{GL}(d, \mathbb{R})$. However, as pointed out by Viana [47, page 678], this submersion lemma requires that $\dim G \geq n(n-1)$. This brings us to the question of understanding which groups can be led to obtain a non-trivial spectrum. For example, the symplectic group $\mathrm{Sp}(2n, \mathbb{R})$ when $n \geq 2$ is outside the scope of Viana's lemma and so a different approach was done in [21].

Due to algebraic and dimensional issues our kinetic setting is not covered by the submersion lemma and so a different kind of strategy must be used to get similar results. The question of knowing the C^r-open and dense prevalence of simple Lyapunov spectrum arising from equations like (2) is a work in progress. We may ask the following two questions:

Open Question 3 *Kinetic cocycles Φ^t_A display at least one non-zero Lyapunov exponent at μ-almost every point when we take A in an open and dense set of maps in $C^{r,\nu}(M, \mathcal{K})$?*

Open Question 4 *Traceless kinetic cocycles Φ^t_A display non-zero Lyapunov exponents at μ-almost every point when we take A in an open and dense set of maps in $C_0^{r,\nu}(M, \mathcal{K})$?*

6 What About Third Order Linear Homogeneous Differential Equations?

6.1 Jerk Equations

Now we consider the generalization of the case handled so far when we consider Eq. (2). Consider a third order linear homogeneous differential equation

$$\dddot{x}(t) + \alpha(\varphi^t(\omega))\ddot{x}(t) + \beta(\varphi^t(\omega))\dot{x}(t) + \gamma(\varphi^t(\omega))x(t) = 0, \tag{15}$$

where $\alpha, \beta, \gamma\colon M \to \mathbb{R}$ are continuous functions. Let $C^0(M, \mathcal{K}) \subset C^0(M, \mathbb{R}^{3\times 3})$ stand for the subset of *kinetic infinitesimal generators* related with (15) which will be defined by

$$A(w) = \begin{pmatrix} 0 & 1 & 0 \\ 0 & 0 & 1 \\ -\gamma(w) & -\beta(w) & -\alpha(\omega) \end{pmatrix}. \tag{16}$$

These third order differential equations are called *jerk equations*. From the Liouville-Ostrogradski formula and since $\Phi_A^0(\omega) = Id$ we get $\det \Phi_A^t(\omega) = 1$ when $\alpha = 0$ realizing why traceless systems give rise to conservative solutions. In this section we will be interested in the traceless kinetic linear differential systems i.e. in (16) take $\alpha = 0$. Let $C_0^0(M, \mathcal{K}) \subset C^0(M, \mathcal{K})$ be[2] the subset of kinetic traceless linear differential systems. Clearly, $C_0^0(M, \mathcal{K})$ is a topologically closed set in both $C^0(M, \mathcal{K})$ and $C^0(M, \mathbb{R}^{3\times3})$ with respect to the C^0-topology.

6.2 Partial Hyperbolicity

We say that $E = E^u \oplus E^c \oplus E^s$ is a (C, σ)-(uniformly) *partially hyperbolic* for Φ_A^t if it is Φ_A^t-invariant, E^u is hyperbolic expanding and E^s is hyperbolic contracting; moreover these two subbundles are not simultaneously trivial. We have that E^u dominates E^c and E^c dominates E^s. Recall that by Oseledets theorem we get:

$$\lim_{t \to \pm\infty} \frac{1}{t} \log|\det(\Phi_A^t)| = \lambda^s(A) + \lambda^c(A) + \lambda^u(A) = 0. \tag{17}$$

6.3 Removing Zero Lyapunov Exponents on a Certain Jerk Equation

Assuming that we have a proper partial hyperbolic decomposition $E^s \oplus E^c \oplus E^u$ displaying a zero Lyapunov exponent along the central direction E^c, in [18] was proved the next result that some C^0 perturbation of the parameters $\beta(t)$ and $\gamma(t)$ can be done in order to obtain non-zero Lyapunov exponents and so a chaotic behaviour of the solution.

Theorem 3 ([18]) *Let $\varphi^t : M \to M$ be an ergodic flow w.r.t. a probability volume measure, $A \in C_0^0(M, \mathcal{K})$ and assume that the cocycle Φ_A^t has a partially hyperbolic splitting $E^u \oplus E^c \oplus E^s$ over M. Then, either $\lambda^c(A) \neq 0$, or else A may be approximated, in the C^0-topology, by $A_0 \in C^0(M, \mathcal{K}_0)$ for which $\lambda^c(A_0) \neq 0$.*

The sketch of the proof of Theorem 3 (based on [12, 19]) is the following. For full details see [18].

[2] We use the same notation as in Sect. 4 but of course that now we are in $\mathbb{R}^{3\times3}$.

Step 1 We first perturb locally and kinetically but require that the stable direction remains unchanged;

Step 2 For this to happen choose a perturbation $B = A + P$ such that P act as a null vector field along the stable direction;

Step 3 By the two previous points we sent to the space E^{cu} the whole issue about a balanced increase/decrease of Lyapunov exponents;

Step 4 Using the hypothesis that the system is *conservative*[3] the quantity lost (or gained) in the unstable Lyapunov exponent is equal to the quantity gained (or lost) in the central Lyapunov exponent;

Step 5 Finally, since the base flow is ergodic a single perturbation ensures a non-zero central Lyapunov exponent along a certain orbit which is enough.

Acknowledgments The author was partially supported by CMUP, which is financed by national funds through FCT-Fundação para a Ciência e a Tecnologia, I.P., under the project with reference UIDB/00144/2020 and also partially supported by the Project 'Means and Extremes in Dynamical Systems' (PTDC/MAT-PUR/4048/2021).

References

1. W. Ambrose, S. Kakutani, *Structure and continuity of measure preserving transformations*, Duke Math. J., 9, (1942) 25–42.
2. D. Amaro, M. Bessa, H. Vilarinho, *Genericity of trivial Lyapunov spectrum for L^p-cocycles derived from second order linear homogeneous differential equations* ArXiv 2023. arXiv:2301.04905.
3. D. Amaro, M. Bessa, H. Vilarinho, *Simple Lyapunov spectrum for linear homogeneous differential equations with L^p parameters* ArXiv 2023. arXiv:2301.04909.
4. A. Arbieto, J. Bochi, *L^p-generic cocycles have one-point Lyapunov spectrum*, Stochastics and Dynamics 3 (2003) 73–81. Corrigendum. ibid, 3 (2003) 419–420.
5. L. Arnold, *Random Dynamical Systems*, Springer Verlag, 1998.
6. L. Arnold, N. Cong, *Linear cocycles with simple Lyapunov spectrum are dense in L^∞*, Ergod. Th. & Dynam. Sys., 19 (1999) 1389–1404.
7. L. Arnold, N. Cong, *Generic properties of Lyapunov exponents*, Random Comput. Dynam. 2(3-4) (1994) 335–345.
8. L. Arnold, N. Cong, *On the simplicity of the Lyapunov spectrum of products of random matrices*, Ergod. Th. & Dynam. Sys. 17 (1997) 1005–1025.
9. L. Arnold, H. Crauel, J.-P. Eckmann, editors *Lyapunov Exponents.* Proceedings, Oberwolfach 1990, volume 1486 of Springer Lecture Notes in Math. Springer-Verlag, Berlin Heidelberg New York, 1991.
10. L. Arnold, V. Wihstutz, editors, *Lyapunov Exponents.* Proceedings, Bremen 1984, volume 1186 of Springer Lecture Notes in Mathematics. Springer Verlag, Berlin Heidelberg New York, 1986.
11. A. Avila, *Density of positive Lyapunov exponents for $SL(2, \mathbb{R})$-cocycles*, J. Amer. Math. Soc. 24 (4) (2011) 999–1014.

[3] In this case $\lambda^c(A) + \lambda^u(A) = \lambda^c(B) + \lambda^u(B)$ since by Step 1 we have $\lambda^s(A) = \lambda^s(B)$ (recall (17)).

12. A. Baraviera, C. Bonatti, *Removing zero Lyapunov exponents*, Ergod. Th. & Dynam. Sys., 23, (2003) 1655–1670.
13. L. Barreira, Ya. Pesin, Nonuniform Hyperbolicity, Encyclopedia of Mathematics and Its Applications 115, Cambridge University Press, 2007.
14. M. Bessa, *Dynamics of generic 2-dimensional linear differential systems*, J. Differential Equations 228 (2) (2006) 685–706.
15. M. Bessa, *The Lyapunov exponents of generic zero divergence 3-dimensional vector fields*. Ergod. Th. & Dynam. Sys. 27 (2007) 1445–1472.
16. M. Bessa, *Dynamic of generic multidimensional linear differential systems*, Advanced Nonlinear Studies 8 (2008) 191–211.
17. M. Bessa, *Perturbations of Mathieu equations with parametric excitation of large period* Advances in Dynamical Systems and Applications, 7 (1) (2012) 17–30.
18. M. Bessa, *Plenty of hyperbolicity on a class of linear homogeneous jerk differential equations*, Aequationes mathematicae, 97 (2023) 467–487.
19. M. Bessa, J. Rocha, *Removing zero Lyapunov exponents in volume-preserving flows*, Nonlinearity 20 (2007) 1007–1016.
20. M. Bessa, H. Vilarinho, *Fine properties of L^p-cocycles which allows abundance of simple and trivial spectrum*. J. Differential Equations, 256 (7) (2014) 2337–2367.
21. M. Bessa, J. Bochi, M. Cambrainha, C. Matheus, P. Varandas, D. Xu, *Positivity of the Top Lyapunov Exponent for Cocycles on Semisimple Lie Groups over Hyperbolic Bases*, Bull Braz Math Soc, New Series 49 (2018) 73–87.
22. J. Bochi, *Genericity of zero Lyapunov exponents*, Ergod. Th. & Dynam. Sys. 22 (2002) 1667–1696.
23. J. Bochi, M.Viana, *The Lyapunov exponents of generic volume-preserving and symplectic maps*, Ann. of Math. 161 (3) (2005) 1423–1485.
24. Bonatti, C., Gómez-Mont, X., Viana, M., *Généricité d'exposants de Lyapunov non-nuls pour des produits déterministes de matrices*. Ann. Inst. H. Poincaré Anal. Non Linéaire 20 (2003) 579–624.
25. Bonatti, C., Viana, M.: *Lyapunov exponents with multiplicity 1 for deterministic products of matrices*. Ergod. Th. & Dynam. Sys. 24 (2004) 1295–1330.
26. E. Cornelis, M. Wojtkowski, *A criterion for the positivity of the Liapunov characteristic exponent*, Ergod. Theory & Dyn. Syst. 4 (1984) 527–539.
27. N. D. Cong, *A generic bounded linear cocycle has simple Lyapunov spectrum*, Ergod. Th. & Dynam. Sys. 25 (2005) 1775–1797.
28. Duarte, P., Klein, S., *Positive Lyapunov exponents for higher dimensional quasiperiodic cocycles*. Commun. Math. Phys. 332 (1) (2014) 189–219.
29. R. Fabbri, *Genericity of hyperbolicity in linear differential systems of dimension two*, (Italian) Boll. Unione Mat. Ital., Sez. A, Mat. Soc. Cult. 8 (1) Suppl. (1998) 109–111.
30. R. Fabbri, On the Lyapunov Exponent and Exponential Dichotomy for the Quasi-Periodic Schrödinger Operator, Boll. Unione Mat. Ital. 8 (5-B) (2002) 149–161.
31. R. Fabbri, R. Johnson, *Genericity of exponential dichotomy for two-dimensional differential systems*, Ann. Mat. Pura Appl. IV. Ser. 178 (2000) 175–193.
32. R. Fabbri, R. Johnson, *On the Lyapunov exponent of certain $SL(2,\mathbb{R})$-valued cocycles*, Differ. Equ. Dyn. Syst. 7 (3) (1999) 349–370.
33. R. Fabbri, R. Johnson, L. Zampogni, *On the Lyapunov exponent of certain $SL(2,\mathbb{R})$-valued cocycles II*, Differ. Equ. Dyn. Syst. 18 (1-2) (2010) 135–161.
34. X. Feng, K. Loparo, *Almost sure instability of the random harmonic oscillator*, SIAM J. Appl. Math. 50 (3) (1990) 744–759.
35. Ledrappier, F.: *Positivity of the exponent for stationary sequences of matrices*. In: Arnold, L., Wihstutz, V. (eds.) Lyapunov Exponents (Bremen, 1984). Lecture Notes in Mathematics, vol. 1886, pp. 56–73, Springer, New York (1986)
36. T. Kato, Perturbation Theory for Linear Operators, 2nd ed., Springer, 1980.
37. A. Leizarowitz, *On the Lyapunov exponent of a harmonic oscillator driven by a finite-state Markov process*, SIAM J. Appl. Math., 49 (2) (1989) 404–419.

38. R. Mañé, *Oseledec's theorem from generic viewpoint*, Proceedings of the international Congress of Mathematicians, Warszawa vol. 2 (1983) 1259–1276.
39. R. Mañé, *The Lyapunov exponents of generic area preserving diffeomorphisms*, International Conference on Dynamical Systems (Montevideo, 1995), Pitman Res. Notes Math. Ser. 362 (1996) 110–119.
40. V. M. Millionshchikov, *Systems with integral separateness which are dense in the set of all linear systems of differential equations*, Differential Equations 5 (1969) 850–852.
41. M. Nerurkar, *Positive exponents for a dense set of continuous cocycles which arise as solutions to strongly accessible linear differential systems,* Contemp. Math. Ser. AMS 215 (1998) 265–278.
42. O. Knill, *Positive Lyapunov exponents for a dense set of bounded measurable* SL(2, *R*) *cocycles*, Ergod. Th. & Dynam. Sys.. Systems 12 (2) (1992) 319–331.
43. V. L. Novikov, *Almost reducible systems with almost periodic coefficients*, Mat. Zametki 16 (1974) 789–799.
44. V. Oseledets, *A multiplicative ergodic theorem: Lyapunov characteristic numbers for dynamical systems*, Transl. Moscow Math. Soc. 19 (1968) 197–231.
45. R. H. Risch, *The problem of integration in finite terms*, Trans. Amer. Math. Soc. 139 (1969), 167–189.
46. D. Rudolph, *A Two-Valued Step Coding for Ergodic Flows*, Math. Z. 150 (1976) 201–220.
47. M. Viana, *Almost all cocycles over any hyperbolic system have nonvanishing Lyapunov exponents*, Ann. of Math. 167 (2) (2008) 643–680.
48. D. Xu, Density of positive Lyapunov exponents for symplectic cocycles, J. Eur. Math. Soc., 21 (10) (2019), 3143–3190.

An Invitation to $SL_2(\mathbb{R})$ Cocycles Over Random Dynamics

Jamerson Bezerra and Mauricio Poletti

Abstract The purpose of these notes is to discuss the advances in the theory of Lyapunov exponents of linear $SL_2(\mathbb{R})$ cocycles over hyperbolic maps. The main focus is around results regarding the positivity of the Lyapunov exponent and the regularity of this function with respect to the underlying data.

1 Continuous Random Cocycles

We start by introducing the context that will be the used throughout these notes and later, we move on to specific models.

1.1 Continuous Random Cocycles and Lyapunov Exponent

Let $\Sigma = \{1, \dots, \kappa\}^{\mathbb{Z}}$ be the space of infinite bilateral sequences in the symbols $\{1, \dots, \kappa\}$ and let $\sigma : \Sigma \to \Sigma$ be the shift map given by $\sigma((x_n)_n) = (x_{n+1})_n$. Joint with the invariant measure $\mu = p^{\mathbb{Z}}$, $p = (p_1, \dots, p_\kappa)$ being a probability vector, $p_i > 0$, the ergodic system (Σ, σ, μ) will be called *base dynamics*.

Each continuous matrix valued map $A : \Sigma \to SL_2(\mathbb{R})$ uniquely determines a skew-product $F_A : \Sigma \times \mathbb{R}^2 \to \Sigma \times \mathbb{R}^2$ given by $F_A(x, v) = (\sigma(x), A(x)\, v)$. F_A is usually referred to as *Linear cocycle* associated with the base dynamics (Σ, σ, μ) and the fiber action A.

Fixed the base dynamics, we use the term linear cocycle to refer, not only to the map F_A but also to the fiber action A. It is also worth noticing that each map A

J. Bezerra (✉)
Faculty of Mathematics and Computer Science, Nicolaus Copernicus University, Toruń, Poland
e-mail: jdbezerra@mat.umk.pl

M. Poletti
Departamento de Matemática, Universidade do Ceará, Fortaleza, Brasil
e-mail: mpoletti@mat.ufc.br

© The Author(s), under exclusive license to Springer Nature Switzerland AG 2023
J. L. Dias et al. (eds.), *New Trends in Lyapunov Exponents*, CIM Series in Mathematical Sciences, https://doi.org/10.1007/978-3-031-41316-2_2

induces in a standard way a skew-product action on $\Sigma \times \mathbb{P}^1$ that we still denote by F_A.

We also use the following convenient notation (motivated by the chain rule for derivatives):

$$A^n(x) = \begin{cases} A(f^{n-1}(x)) \cdots A(f(x))A(x) & \text{if } n > 0 \\ I & \text{if } n = 0 \\ A(f^n(x))^{-1} \cdots A(f^{-2}(x))^{-1} A(f^{-1}(x))^{-1} & \text{if } n < 0. \end{cases}$$

The Lyapunov exponent associated with the map A may be defined as the exponential growth rate of the quantities $\|A^n(x)\|$. More specifically, by Kingman's sub-additive ergodic theory, for μ-a.e. $x \in \Sigma$, the limit

$$L(A) := \lim_{n \to \pm\infty} \frac{1}{n} \log \left\| A^n(x) \right\|,$$

exists and is constant. The number $L(A)$ is called the *Lyapunov exponent* of the linear cocycle A. This quantity measures the amount of hyperbolicity produced by the fiber action A along the orbits of μ typical points in the basis and so it can be used as a measure of the chaoticity of the system expressed by the skew-product F_A. For that reason, the following two problems becomes central in the theory.

1. **Positivity problem:** How big is the set of A with positive Lyapunov exponent?
2. **Regularity problem:** How regular is the map $A \mapsto L(A)$?

Disclaimer Lyapunov exponents for linear cocycles are known in a much more general framework. For instance, we could consider more general basis dynamics or even higher dimensional fiber actions. However, we restrict ourselves to the case of Bernoulli shift and $SL_2(\mathbb{R})$ cocycles for a few reasons. First, we focus on making an exposition of the results that relies on the hyperbolic structure of the base dynamics and the model that best represents such a hyperbolic structure is the full shift (Σ, σ, μ). Second, the results for higher dimensional fiber actions are more scarce and among most of those the core idea in the study can be demystified when we restrict ourselves to the two dimensional case. These restrictions are the authors choices and do not reflect any particular extra importance of the chosen exposition.

Before we proceed, we list a few basic properties of the Lyapunov exponent:

Proposition 1.1 *Consider a continuous map* $A : \Sigma \to SL_2(\mathbb{R})$. *Then*

1. $L(A) \geq 0$;
2. $L(A) = \inf_{n \geq 1} \int_\Sigma \frac{1}{n} \log \|A^n(x)\| \, d\mu(x)$;
3. *The map* $A \mapsto L(A)$ *is upper semi-continuous. In particular, the maps* A *with zero exponent are continuity points of the function* L;

4. *(Oseledet's theorem, [55]) If $L(A) > 0$, then for μ-a.e. $x \in \Sigma$, there exist A-invariant projective directions $\hat{e}^u(x), \hat{e}^s(x) \in \mathbb{P}^1$, such that*

$$\lim_{n\to\infty} \frac{1}{n} \log \|A^n(x)\, v\| = \begin{cases} L(A), & v \notin \hat{e}^s(x) \\ -L(A), & v \in \hat{e}^s(x). \end{cases}$$

$$\lim_{n\to\infty} \frac{1}{n} \log \|A^{-n}(x)\, v\| = \begin{cases} -L(A), & v \notin \hat{e}^u(x) \\ L(A), & v \in \hat{e}^u(x). \end{cases}$$

Moreover, the invariant sections in $\Sigma \times \mathbb{P}^1$, $x \mapsto \hat{e}^s(x), \hat{e}^u(x) \in \mathbb{P}^1$ are measurable;

5. *The direction $\hat{e}^s(x)$ ($\hat{e}^u(x)$) only depends on the positive (negative) coordinates of x.*

Remark 1.1 (Notation) We use the notation $\hat{v}$ to denote elements of the projective space $\mathbb{P}^1$. If $\hat{v}$ and v appears in the same expression, v is (one) unitary vector in $\mathbb{R}^2$ in the direction determined by $\hat{v}$.

Outline of the Proof Item 1, is obtained using the fact that A takes values in $SL_2(\mathbb{R})$. Item 2, is due to the characterization of the value of the Lyapunov exponent given by Kingman's sub-additive ergodic theorem. Item 3, is a direct consequence of Item 2. The item 4 is the Oseledets Theorem and item 5 comes from the proof of Oseledets theorem using that

$$\hat{e}^s(x) = \lim \hat{s}(A^n(x)) \quad \text{and} \quad \hat{e}^u(x) = \lim A^n(\sigma^{-n}(x))\, \hat{u}(A^n(\sigma^{-n}(x))),$$

where for a matrix $B \in SL_2(\mathbb{R})$ with $\|B\| > 1$, $\hat{u}(B)$ and $\hat{s}(B)$, denotes the singular directions associated to B associated respectively with the biggest and smallest singular values. □

1.2 Uniform Hyperbolicity

Item 3 of the Proposition 1.1 is the start point of the study of the soft regularity of the Lyapunov exponent function. For cocycles with positive Lyapunov exponent, however, the regularity of L may have a very nasty behavior. In order to properly discuss this matter it is essential to introduce the class of uniformly hyperbolic cocycles.

We say that A is *uniformly hyperbolic*, if there exist continuous A-invariant sections $\Sigma \ni x \mapsto \hat{e}^u(x), \hat{e}^s(x) \in \mathbb{P}^1$ (defined everywhere in Σ), and constants $C, \lambda > 0$, such that for every $x \in \Sigma$,

$$\|A^n(x)|_{\hat{e}^s(x)}\| \le C\, e^{-\lambda n} \quad \text{and} \quad \|A^{-n}(x)|_{\hat{e}^u(x)}\| \le C\, e^{-\lambda n}.$$

For a matrix $B \in SL_2(\mathbb{R})$ and a direction $\hat{v} \in \mathbb{P}^1$, $B|_{\hat{v}}$ denotes the restriction of B to the subspace determined by $\hat{v}$. Every uniformly hyperbolic cocycle A has positive Lyapunov exponent. Indeed, using the definition we see that $L(A) \geq \lambda > 0$.

Remark 1.2 The directions $\hat{e}^u$, $\hat{e}^s$, defined above, coincide with the directions given by Oseledets Theorem (see Item 4 of Proposition 1.1).

An alternative characterization of the uniform hyperbolicity for cocycles comes from the *cone field criteria*: for every $x \in \Sigma$ there exists a pair of disjoint closed projective intervals, $\hat{J}^s(x)$, $\hat{J}^u(x) \subset \mathbb{P}^1$, such that

$$A(x)(\hat{J}^u(x)) \subset \mathrm{Int}(\hat{J}^u(\sigma(x))) \quad \text{and} \quad A^{-1}(x)(\hat{J}^s(x)) \subset \mathrm{Int}(\hat{J}^s(\sigma(x)))$$

(see Fig. 1). Using the cone field criteria, it is clear that uniform hyperbolicity is a stable property among the maps $A \in C^0(\Sigma, SL_2(\mathbb{R}))$.

Remark 1.3 For $SL_2(\mathbb{R})$ cocycles a more handy criterion to determine if a cocycle is uniformly hyperbolic is available: it is enough to find constants $C, \lambda > 0$ satisfying that for every $x \in \Sigma$ and $n \geq 1$,

$$\left\| A^n(x) \right\| \geq C\, e^{\lambda n}.$$

For a proof see [69].

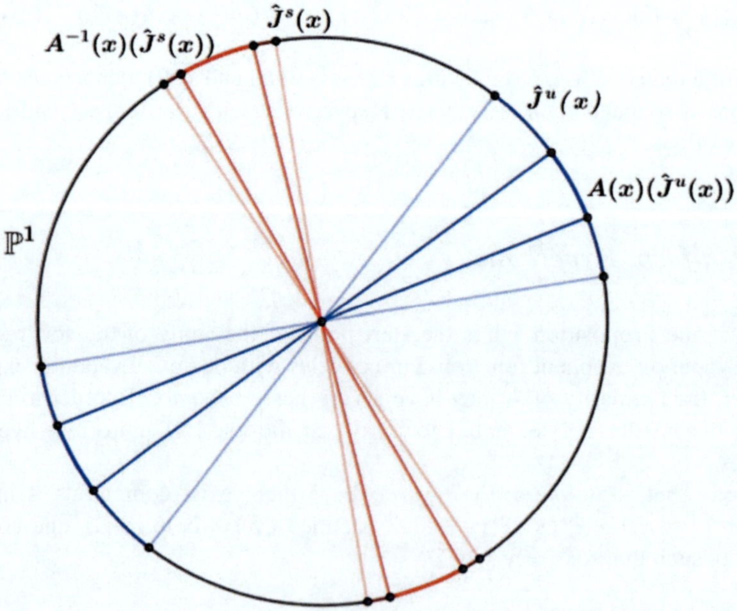

Fig. 1 Projective cone field

Example 1.1 (Positive Cocycles) Consider $A : \Sigma \to SL_2(\mathbb{R})$ continuous such that for all $x \in \Sigma$ all of the entries of the matrix $A(x)$ are strictly positive. Then, A is uniformly hyperbolic. Indeed, A preserves the constant cone field formed by I ($\mathbb{P}^1 \backslash I$ is preserved by the backwards action), with I being the projectivization of the first quadrant in $\mathbb{R}^2$ i.e., $\{(s,t) \in \mathbb{R}^2 : s, t \geq 0\}$.

Example 1.2 (Triangular Cocycles) Consider continuous functions $a, b, d : \Sigma \to \mathbb{R}$, with $\zeta := \inf_\Sigma |a| > 1$, and define $A : \Sigma \to SL_2(\mathbb{R})$ by $A(x) = \begin{pmatrix} a(x) & b(x) \\ 0 & d(x) \end{pmatrix}$. then $\|A^n(x)\| \geq \zeta^n$, for every $x \in \Sigma$ and every $n \geq 1$. So, using the equivalent definition mentioned in Remark 1.3, we see that A is uniformly hyperbolic.

Example 1.3 (Schrödinger Cocycles) Another family of example of uniformly hyperbolic cocycles is provided by the so called *Schrödinger cocycles* that we describe now. Let $\varphi : \Sigma \to \mathbb{R}$ be a continuous function. Consider the family of self-adjoint bounded operators $\{H_x : \ell^2(\mathbb{Z}) \to \ell^2(\mathbb{Z})\}_{x \in \Sigma}$ given by

$$(H_x\, u)_n := (-\Delta u + \Phi_x \cdot u)_n = -u_{n+1} - u_{n-1} + \varphi(\sigma^n(x))\, u_n.$$

The operator H_x, for each $x \in \Sigma$, is called the (discrete) *dynamically defined Schrödinger operator* associated with the potential function φ at the point $x \in \Sigma$.

The ergodicity of our base dynamics implies that the spectrum of the operator H_x does not depend on the choice of point x for μ-a.e. $x \in \Sigma$ (see [28]). In a naive attempt to solve the eigenvalue-eigenvector equation, for a vector $u = (u_n)_n$ and $E \in \mathbb{R}$, we end up with a second order linear recursive equation that may be described in a matrix form as follows: for every $n \geq 0$,

$$\begin{pmatrix} \varphi(\sigma^{n-1}(x)) - E & -1 \\ 1 & 0 \end{pmatrix} \cdots \begin{pmatrix} \varphi(x) - E & -1 \\ 1 & 0 \end{pmatrix} \begin{pmatrix} u_0 \\ u_{-1} \end{pmatrix} = \begin{pmatrix} u_n \\ u_{n-1} \end{pmatrix}. \tag{1}$$

The *Schrödinger cocycle* $A_E : \Sigma \to SL_2(\mathbb{R})$ associated to the *energy* $E \in \mathbb{R}$ is defined by

$$A_E(x) = \begin{pmatrix} \varphi(x) - E & -1 \\ 1 & 0 \end{pmatrix}.$$

Notice that the right-hand side of the Eq. (1) described the orbit of the vector $\bar{u} = (u_0, u_{-1}) \in \mathbb{R}^2$ by the cocycle A_E. This action produces a formal eigenvector $u = (u_n)_n$ associated to the eigenvalue $E \in \mathbb{R}$. However, any growth rate of the sequence $n \mapsto \left\| A_E^n(x)\, \bar{u} \right\|$ is incompatible with the chance of $u \in \ell^2(\mathbb{Z})$. This is the first indication of the close relationship of study of Schrödinger spectrum and the Lyapunov exponent of the Schrödinger cocycle A_E as a function of the energy E. This close relation is confirmed by the next result.

Proposition 1.2 (Johnson [45]) *The real number E is not in the spectrum of the Schrödinger operator $\{H_x\}_{x\in\Sigma}$ if and only if A_E is uniformly hyperbolic.*

Using the cone field criteria, it is not hard to see that the set of uniformly hyperbolic cocycles form an open set inside of the set of continuous functions $C^0(\Sigma, SL_2(\mathbb{R}))$ endowed with the sup norm. Back to problems 2.1 and 2.1, we see directly from the definition, that if A is uniformly hyperbolic, then $L(A) > 0$. Actually, in this set of cocycles we have the best regularity that we can expect. That is the content of the next result.

Theorem 1.1 (Ruelle [61]) *The Lyapunov exponent function L restricted to the set of uniformly hyperbolic cocycles is a real analytic function.*

Ruelle obtained asymptotic formulas for the derivative of the Lyapunov exponent in terms of the invariant directions of the uniformly hyperbolic cocycle. Therefore, the regularity of the Lyapunov exponent is a consequence of the nice behaviour of the invariant directions with respect to the cocycle due to the contraction properties (cone contractions) provided by the uniform hyperbolicity.

1.3 Regularity for Continuous Cocycles

So far, we have seen that cocycles with zero Lyapunov exponent are continuity points of L and we have also discussed the regularity among the open class of uniformly hyperbolic cocycles. Now, we move forward to analyze what happens in the complement of these two sets. In other words, what can be said about cocycles that have positive Lyapunov exponent but fail to be uniformly hyperbolic? See Example 3.1. This issue was addressed first by R. Mañe and later formalized by J. Bochi.

Theorem 1.2 (Bochi [17]) *If $A \in C^0(\Sigma, SL_2(\mathbb{R}))$ is not uniformly hyperbolic and $L(A) > 0$, then there exists a sequence $(A_n)_n \subset C^0(\Sigma, SL_2(\mathbb{R}))$ converging to A such that $L(A_n) = 0$.*

As a consequence, we have the following dichotomy: for cocycles A with non-vanish Lyapunov exponent, either A is uniformly hyperbolic and L is analytic in a neighborhood of A, or A can be approximated by cocycles with zero Lyapunov exponent. Notice that Theorem 1.2 finishes the Problem 2.1 for cocycles in $C^0(\Sigma, SL_2(\mathbb{R}))$.

Theorem 1.2 relies on the flexibility to design local perturbations provided by the C^0-topology. This is essential in the proof and the strategy goes as follows: the fact that the Lyapunov exponent of A is positive guarantees the existence of the measurable Oseledets sections $\hat{e}^s$, $\hat{e}^u$ such as in item 3 of Proposition 1.1. However, the fact that the cocycle is not uniformly hyperbolic implies that up to small C^0 local perturbations the distance between $\hat{e}^s$ and $\hat{e}^u$ is arbitrarily close to zero in positive measure sets (if it is bounded away from zero, it would be possible to build cone fields).

Once we have that the distance between $\hat{e}^s(x)$ and $\hat{e}^u(x)$ is small, we may, again, perform a local C^0-perturbation (which can not be achieved in higher regularity) of the cocycle A obtaining a new cocycle B with $B(x) \cdot \hat{e}^s(x) = \hat{e}^u(x)$ for this set of points. Changing the Oseledets directions kills the exponential growth rate of the norms $\|B^n(x)\|$ which is enough to guarantee that the Lyapunov exponent vanishes.

Remark 1.4 Formally speaking, the procedure described above could be occurring over a coboundary set. In this situation, the Oseledets directions could be swapped keeping the positivity of the Lyapunov exponent. But this technical issue can always be solved in aperiodic system such as our base dynamics. See [69, Section 9.2.2] for a precise discussion of this issue.

The previous strategy, however, does not work if we try to perform the same type of perturbation in higher regularity C^γ with $\gamma > 0$. Here, the size of the support of a local perturbation must to be related with the amount that we are allowed to perturb. The regularity of the Lyapunov exponent function is still open in the C^γ-topology. We will come back to this discussion later in Sect. 3.

1.4 Continuous Cocycles with Positive Lyapunov Exponent

Before addressing the regularity of the Lyapunov exponent associated with more restrictive topologies, we highlight that a byproduct of Theorem 1.2 is the fact that, in $C^0(\Sigma, SL_2(\mathbb{R}))$, outside of uniformly hyperbolic cocycles, there is no open set formed only by positive Lyapunov exponent. However, we can still have plenty of points with that property. This is exactly the content of the next result.

Theorem 1.3 (Avila [1]) *The set* $\{A \in C^\gamma(\Sigma, SL_2(\mathbb{R})) \colon L(A) > 0\}$ *is dense in* $C^\gamma(\Sigma, SL_2(\mathbb{R}))$ *for any* $\gamma \geq 0$.

The proof of this result is a consequence of the so called *regularization expressions* for the Lyapunov exponent, the simplest of which is a (global) expression provided in [2],

$$\int_0^1 L(A R_{2\pi\theta})\, d\theta = \int_\Sigma \log\left(\frac{\|A\| + \|A^{-1}\|}{2}\right) d\mu, \tag{2}$$

where $R_{2\pi\theta}$ is the rigid rotation of angle $2\pi\theta$. In order to obtain Theorem 1.3, local regularizing expressions such as (2) are used to guarantee the positivity of the Lyapunov exponent under small perturbations.

An adaption of the proof of Theorem 1.3 is provided in [1] to guarantee positivity of the Lyapunov exponent with respect of the energy E for Schrödinger cocycles $A_{E,\varphi}$ (see Example 1.3). More precisely, the result says that for a dense set of potentials $\varphi \in C^0(\Sigma)$, we have $L(A_{E,\varphi}) > 0$ for a dense set of energies $E \in \mathbb{R}$.

Remark 1.5 Up to now, all of the mentioned results do not use the hyperbolic structure of the base dynamics (Σ, σ, μ).

2 Locally Constant Cocycles

To understand the properties resulting from the hyperbolic structure of the base dynamics, we focus our attention on a specific finite dimensional subspace of $C^0(\Sigma, SL_2(\mathbb{R}))$ in which the base dynamics has a direct influence on the fiber action. These cocycles, known as *locally constant cocycles*, are maps $A : \Sigma \to SL_2(\mathbb{R})$ which are constant in cylinders of the form $[0; i] := \{x \in \Sigma \colon x_0 = i\}$, $i = 1, \ldots, \kappa$.

The theory of Lyapunov exponents of locally constant cocycles has been actively developed over the past 60 years. Because many of the main ideas used in modern techniques to study the regularity and positivity of the Lyapunov exponents come from the locally constant setting, it is worth discussing it in a bit more detail.

2.1 *Random Product of Matrices*

In order to properly introduce this probabilistic setup, let $\{A_1, \ldots, A_\kappa\}$ be a finite set of matrices in $SL_2(\mathbb{R})$. Consider a sequence of independent random matrices $L_1, \ldots, L_n, \ldots$ on $SL_2(\mathbb{R})$ with a common Bernoulli distribution given by the probability measure $p_1\delta_{A_1} + \ldots + p_\kappa\delta_{A_\kappa}$, associated to the data (p, A), where $p = (p_1, \ldots, p_\kappa)$ and $A = (A_1, \ldots, A_\kappa) \in SL_2(\mathbb{R})^\kappa$. Furstenberg and Kesten in [38], obtained that the limit

$$L(p, A) = \lim_{n\to\infty} \frac{1}{n} \log \|L_n \cdot \ldots \cdot L_1\| .$$

exists p-almost surely. The quantity $L(p, A)$ is called the Lyapunov exponent of the random sequence $L_1, \ldots, L_n, \ldots$. It is also called the *Lyapunov exponent of the product of the random matrices* $(A_1, \ldots, A_\kappa)$. This Lyapunov exponent only depends on the data (p, A). So, the focus of the study is to analyze the properties of the real function which associate the data (p, A) to the Lyapunov exponent $L(p, A)$.

Before proceeding with the comparison of this probabilistic model with the previously discussed linear cocycles, it is important to highlight that the dependence of the quantity $L(p, A)$ on the probability vector $p = (p_1, \ldots, p_\kappa)$ is in general more regular than the dependence on the matrix vector $A = (A_1, \ldots, A_\kappa) \in SL_2(\mathbb{R})^\kappa$. In fact, the next result due to Y. Peres shows that the dependence of $L(p, A)$ on the probability vector is highly regular.

Theorem 2.1 (Peres [58]) *If $p_i \neq 0$, for every i and $L(p, A) > 0$, then there exists a neighborhood of p, formed by probability vectors q, such that the function $q \mapsto L(q, A)$ is real analytic.*

The strategy to obtain this result is to study the so called *Markov operator* associated to (p, A). This is defined as the linear operator $Q_p = Q_{(p,A)} : C^0(\mathbb{P}^1) \to C^0(\mathbb{P}^1)$ with

$$Q_p(\varphi)(\hat{v}) := \sum_{i=1}^{\kappa} p_i\, \varphi(A_i\, \hat{v}),$$

(the notation only highlights the dependence on p once A is fixed). The idea of the proof of Theorem 2.1 is that, for every $\hat{v} \in \mathbb{P}^1$, we can extend the maps $F_i := p \mapsto Q_p^n(\log \|A_i\,(\cdot)\|)(\hat{v})$ to complex variable functions $z \mapsto Q_z^n(\log \|A_i\,(\cdot)\|)(\hat{v})$ which are homogeneous polynomials of degree n in the variables of z. Using the fact that $L(p, A) > 0$, its is possible to show that for points z close to p, the sequence of functions

$$z \mapsto Q_z^n(\log \|A_i\,(\cdot)\|)(\hat{v}),$$

converges and hence the limit is a holomorphic function of the variable z. The main tool for the convergence of this sequence is a uniform contraction on average provided by the positivity of the Lyapunov exponent (and a generic assumption, see Step 1 in the proof of Theorem 2.6 below) proved first in [56]: for every sufficiently small $\theta > 0$ there exists $\lambda > 0$ such that for every n large enough, we have

$$\sup_{\hat{u},\hat{v}\in\mathbb{P}^1} \int \left(\frac{d(A^n(x)\,\hat{u},\ A^n(x)\,\hat{u})}{d(\hat{u},\hat{v})}\right)^{\theta} d\mu(x) \leq e^{-\lambda n}. \tag{3}$$

To finish, it is enough to notice that, for each probability vector q close to p we have that the sum $\sum_i Q_q^n(\log \|A_i\,(\cdot)\|)(\hat{v})$ converges to the Lyapunov exponent $L(q, A)$.

From Theorem 2.1, to fully understand the function $(p, A) \mapsto L(p, A)$, we may fix the probability vector p and restrict our attention to the map $A = (A_1, \ldots, A_\kappa) \mapsto L(A)$. This is exactly the dynamical context that we had before, once we make the natural identification of $A = (A_1, \ldots, A_\kappa) \in SL_2(\mathbb{R})^\kappa$ with the locally constant cocycle cocycle $A : \Sigma \to SL_2(\mathbb{R})$ given by $A((x_n)_n) = A_{x_0}$. So, as described in Sect. 1, we are interested in the positivity and regularity of the function that associates each vector of matrices A in the 3κ-dimensional manifold, $SL_2(\mathbb{R})^\kappa$, to the real number $L(A)$.

2.2 *Stationary Measures and Criteria for Positivity of the LE*

An essential tool to understand the Lyapunov exponent of locally constant cocycles are the so called *stationary measures*. These projective probability measures carry all the asymptotic information of the fiber action A and are defined as the fixed points of the dual of the Markov operator Q_A. More precisely, we say that a probability measure η on $\mathbb{P}^1$ is *forward stationary* (or simply stationary) for the cocycle A if $\eta = \sum_{i=1}^{\kappa} p_i (A_i)_* \eta$. We say that η is *backward stationary* if $\sum_{i=1}^{\kappa} p_i (A_i^{-1})_* \eta$

The next proposition provides a few properties of the stationary measures useful for later discussion.

Proposition 2.1 *Let $A \in SL_2(\mathbb{R})^\kappa$ be a locally constant cocycle. Consider Σ^+ the set of positive sequences $(x_n)_{n\geq 0}$ and let $F_A^+ : \Sigma^+ \times \mathbb{P}^1 \to \Sigma^+ \times \mathbb{P}^1$ be the cocycle in $\Sigma^+ \times \mathbb{P}^1$ induced by A.*

1. *The set of stationary measures Stat(A) is non-empty compact and convex;*
2. *It holds that, $\eta \in$ Stat(A) if and only if $\mu \times \eta$ is F_A^+-invariant. Moreover, η is extremal in Stat(A) if and only the system $(F_A^+, \mu \times \eta)$ is ergodic;*
3. *For any extremal stationary measure η we have that*

$$\int_{\Sigma\times\mathbb{P}^1} \log \|A(x)\, v\| \; d(\mu \times \eta)(x, \hat{v}) \in \{L(A), -L(A)\};$$

4. *(Furstenberg's formula): It holds that*

$$L(A) = \sup_{\eta\in Stat(A)} \int_{\Sigma\times\mathbb{P}^1} \log \|A(x)\, v\| \; d(\mu \times \eta)(x, \hat{v}). \tag{4}$$

Outline of the Proof of Proposition 2.1 Item 1 is a classical argument similar to Bogoliouboff's Theorem. The first part of Item 2 is a direct computation and in the second part we use Item 1.

For item 3. Take $\eta \in \mathrm{Stat}(A)$, extremal, and consider the function $\Phi_A : \Sigma\times\mathbb{P}^1 \to \mathbb{R}$ given by $\Phi_A(x, \hat{v}) = \log \|A(x)\, v\|$. Then, by the ergodic theorem,

$$\begin{aligned}\lim_{n\to\infty} \frac{1}{n} \log \left\|A^n(x)\, v\right\| &= \lim_{n\to\infty} \frac{1}{n} \sum_{j=0}^{n-1} \Phi_A((F_A^+)^j(x, \hat{v})) \\ &= \int_{\Sigma\times\mathbb{P}^1} \log \|A(x)\, v\| \; d(\mu \times \eta)(x, \hat{v}),\end{aligned}$$

for $(\mu\times\eta)$-a.e. $(x, \hat{v})$. Using item 4 of Proposition 1.1 and the fact that η is extremal, we see that for μ-a.e. $x \in \Sigma$, the left-hand-side above is either $L(A)$ or $-L(A)$. Item

4, is a consequence of item 3 and the ergodic decomposition for stationary measures. See [69, Section 5] for detailed proofs. □

Remark 2.1 One may wonder why in the item 3 above we need to restrict our attention to the cocycle F_A^+ generated by the *one-sided shift* (Σ^+, σ) and not the usual cocycle F_A. The main reason is because the measure $\mu \times \eta$ is F_A^+-invariant but may not be F_A-invariant.

To see this, take the set $[-1;\ j] \times \hat{V}$. Then, $F_A^{-1}([-1;\ j] \times \hat{V}) = [0;\ j] \times A_j^{-1}(\hat{V})$ and thus, $\mu \times \eta(F_A^{-1}([-1;\ j] \times \hat{V})) = p_j \cdot (A_j)_* \eta(V)$. So, $\mu \times \eta$ is F_A-invariant if and only if $(A_j)_* \eta = \eta$ for every $j = 1, \ldots, \kappa$.

Item 3 and 4 of Proposition 2.1 are very important tools relating the Lyapunov exponent and the stationary measures of the cocycle. This will be used many times in the rest of these notes, starting with the following example:

Example 2.1 (Kifer's Example, [49]) Take $\alpha > 1$ and consider the matrices

$$A_1 = \begin{pmatrix} \alpha & 0 \\ 0 & \alpha^{-1} \end{pmatrix} \quad \text{and} \quad A_2 = R_{\pi/2}.$$

Let A be the locally constant cocycle generated by A_1 and A_2 with probabilities $p_1, p_2 > 0$. The measure $\eta = 1/2\delta_{\hat{e}_1} + 1/2\delta_{\hat{e}_2}$, where e_1 and e_2 denote respectively the unitary horizontal and vertical direction, is a stationary measure for A. Furthermore, a direct computation shows that

$$\int_{\Sigma \times \mathbb{P}^1} \log \|A(x)\, v\|\, d(\mu \times \eta)(x, \hat{v}) = 0.$$

Later (see Proposition 2.2), we are going to see that η is indeed the unique stationary measure for A and so, using Furstenberg's formula (item 4 of Proposition 2.1), $L(A) = 0$.

The last example shows that sometimes we have an easy description of the stationary measures of our cocycle. In general that is not the case, even if we assume that our cocycle has very good properties (with respect to the regularity of the Lyapunov exponent) such as being uniformly hyperbolic:

Example 2.2 (Bernoulli Convolutions) In this example, we fix the probability vector of the base dynamics as $p = (1/2, 1/2)$. Let $\lambda \in (0, 1)$ and set

$$A_1 = \begin{pmatrix} \sqrt{\lambda} & \frac{1}{\sqrt{\lambda}} \\ 0 & \frac{1}{\sqrt{\lambda}} \end{pmatrix} \quad \text{and} \quad A_2 = \begin{pmatrix} \sqrt{\lambda} & -\frac{1}{\sqrt{\lambda}} \\ 0 & \frac{1}{\sqrt{\lambda}} \end{pmatrix}.$$

Let A_λ be the locally constant cocycle generated by A_1 and A_2. For the purpose of this example, we parameterize the projective space $\mathbb{P}^1$ as $\mathbb{R} \cup \{\infty\}$. In these

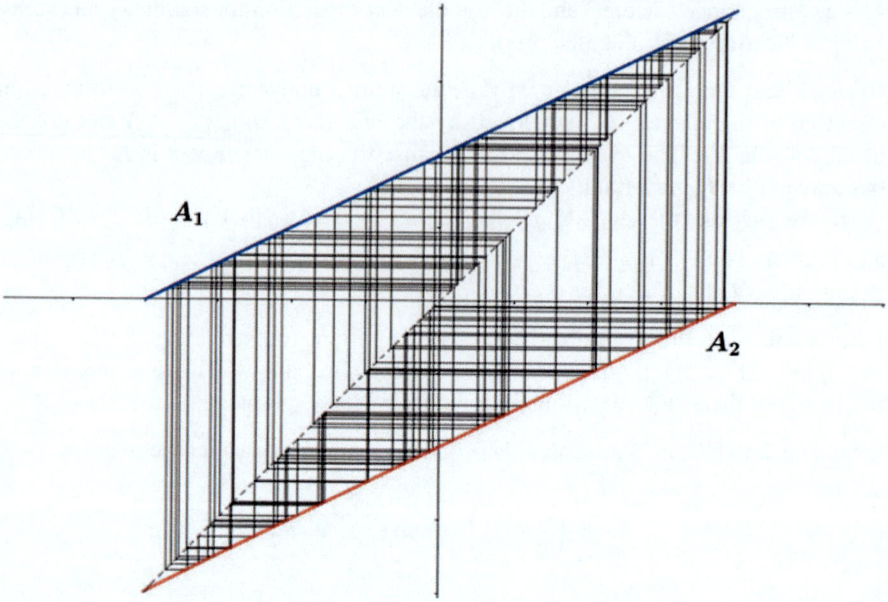

Fig. 2 Random orbit of 0 by $A^j_\lambda(x)(0)$ with $0 \le j \le 100$ and $\lambda = 1/2$

coordinates, we can write the action of a matrix $B = \begin{pmatrix} a & b \\ c & d \end{pmatrix}$ as $B(t) = (at + b)/(ct + d)$. Thus, in these coordinates,

$$A_1(t) = \lambda t + 1 \quad \text{and} \quad A_2(t) = \lambda t - 1.$$

In other words, A_1, A_2 are similarities (see [54, Section 8.3] for a precise definition and properties). See Fig. 2 for a graphical description of the orbit of 0 by this iterated function system.

Let η_λ be the (unique) self-similar measure associated with the affine contractions $A_\lambda := (A_1, A_2)$. Observe that η_λ is supported at $[-1/(1-\lambda), 1/(1-\lambda)]$. It is clear by definition of self similar measure that η_λ is stationary for A_λ. Now, consider the functions $a : \Sigma \to \{+1, -1\}$ and $\tau : \Sigma \to \mathbb{R}$ given by

$$a(x) = \begin{cases} +1, & x_0 = 1 \\ -1, & x_0 = 2. \end{cases} \quad \text{and} \quad \tau(x) = \sum_{j=0}^{\infty} a(\sigma^j(x))\,\lambda^j.$$

We claim that η_λ coincides with the distribution of the random variable τ, $\tau_*\mu$ (recall that μ is the fixed measure on Σ). Indeed, notice that, for any measurable set $B \subset \mathbb{R}$

we have

$$\begin{aligned}\mu(\tau^{-1}(B)) &= \mu\left([0;\ 1] \cap \tau^{-1}(B)\right) + \mu\left([0;2] \cap \tau^{-1}(B)\right)\\ &= \mu\left([0;\ 1] \cap \sigma^{-1} \circ \tau^{-1} \circ (A_1)^{-1}(B)\right)\\ &\quad + \mu\left([0;\ 2] \cap \sigma^{-1} \circ \tau^{-1} \circ (A_2)^{-1}(B)\right)\\ &= \frac{1}{2}(A_1)_*\tau_*\mu(B) + \frac{1}{2}(A_2)_*\tau_*\mu(B).\end{aligned}$$

Hence, $\tau_*\mu$ is a self similar measure supported in $[-1/(1-\lambda), 1/(1-\lambda)]$ and thus by uniqueness of the self similar measure for (A_1, A_2), $\tau_*\mu = \eta_\lambda$. In other words, the stationary measure η_λ is the distribution measure of the Bernoulli convolutions

$$\sum_{j=0}^{\infty} \pm\lambda^j,$$

where the signs $+$ and $-$ are chosen with probability $1/2$.

A classical problem that goes back to Erdös, in [35], is to determine the fractal properties of the measures η_λ. More specifically, the goal is to describe, for each value of λ, if the measure η_λ is singular or absolutely continuous with respect to Lebesgue:

Problem 2.1 *Describe precisely the set of $\lambda \in (0, 1)$ such that η_λ is absolutely continuous or singular with respect to Lebesgue.*

There has been a great progress regarding this problem. See [54] for details about the statements and references therein.

- For $\lambda \in (0, 1/2)$, η_λ is a self similar measure supported on a Cantor set, thus singular with respect to Lebesgue;
- If $\lambda = 1/2$ we have that $\eta_\lambda = \text{Leb}|_{[-2,2]}$;
- For $\lambda \in (1/2, 1)$ the situation is much trickier. Only a countable set of such λ are known where η_λ is singular with respect to Lebesgue. Those are given by the inverse of the so called *Pisot's numbers*: solutions of polynomial equations with integers coefficients such that all other complex solutions have modulus less than one.
- The breakthrough result was provided by B. Solomyak in [64] proving that for almost every $\lambda \in (1/2, 1)$, η_λ is absolutely continuous with respect to Lebesgue.

Notice that for every λ the cocycle A_λ is uniformly hyperbolic. Indeed, it is enough to observe that for every $x \in \Sigma$, $\|A^n(x)\| \geq \lambda^{-n/2}$ and use the criteria provided in 1.3.

To finish the discussion, observe that Dirac measure on the fixed point ∞, δ_∞, is also a stationary measure for A_λ. Hence $\mathrm{Stat}(A_\lambda) = [\eta_\lambda, \delta_\infty]$. Moreover,

$$L(A_\lambda|_{\delta_\infty}) = -L(A) = \frac{1}{2}\log\lambda \quad \text{and} \quad L(A_\lambda|_{\eta_\lambda}) = L(A) = -\frac{1}{2}\log\lambda.$$

Example 2.2 above shows that even among the uniformly hyperbolic cocycles (family in which Lyapunov exponent varies regularly) it is a hard task to understand the properties of the stationary measures.

Nevertheless, the situation is not entirely hopeless. There is a large class of locally constant cocycles A where soft analysis can be used to determine properties of the set $\mathrm{Stat}(A)$. Below, we give a list of concepts which will be useful for our discussion. The concepts are sorted from the strongest to the weakest. We say that the vector (or cocycle) $A = (A_1, \dots, A_\kappa) \in SL_2(\mathbb{R})^\kappa$ is

1. *Strongly-irreducible*, if there is no finite collection of projective directions $\hat{V} = \{\hat{v}_1, \dots, \hat{v}_m\} \subset \mathbb{P}^1$ such that $A_i(\hat{V}) = \hat{V}$ for every $i = 1, \dots, \kappa$;
2. *Irreducible*, if there is no projective direction $\hat{v} \in \mathbb{P}^1$ with $A_i\,\hat{v} = \hat{v}$ for every $i = 1, \dots, \kappa$;
3. *Quasi-irreducible*, if the unique possible direction $\hat{v} \in \mathbb{P}^1$ such that $A_i\hat{v} = \hat{v}$ for every $i = 1, \dots, \kappa$ must to satisfy $L(A|_{\hat{v}}) = L(A)$.

Example 2.3 Here, we mention a few examples satisfying the conditions listed above.

1. *Strongly irreducible:* for any $\beta > 1$ and $\theta \in \mathbb{R}\backslash\mathbb{Q}$,

$$A_1 = \begin{pmatrix} \beta & 0 \\ 0 & \beta^{-1} \end{pmatrix} \quad \text{and} \quad A_2 = R_{2\pi\theta};$$

   Another interesting example is provided by the *Anderson's model.* That is, the family of Schrödinger cocycles A_E over the shift (see Example 1.3) for any energy $E \in \mathbb{R}$, given by

$$A_{1,E} = \begin{pmatrix} a_1 - E & -1 \\ 1 & 0 \end{pmatrix} \quad \text{and} \quad A_{2,E} = \begin{pmatrix} a_2 - E & -1 \\ 1 & 0 \end{pmatrix}.$$

   Assuming $a_1 \neq a_2$, A_E is strongly irreducible for every energy $E \in \mathbb{R}$. For an example with zero Lyapunov exponent is enough to consider the constant cocycle $A \equiv R_{2\pi\theta}$ with $\theta \in \mathbb{R}\backslash\mathbb{Q}$.
2. *Irreducible but not strongly irreducible:* That is provided by Kifer's example (see Example 2.1).

3. *Quasi-irreducible but not irreducible:* consider the triangular cocycle $A \in SL_2(\mathbb{R})^2$ given by

$$A_1 = \begin{pmatrix} a_1 & b_1 \\ 0 & d_1 \end{pmatrix} \quad \text{and} \quad A_2 = \begin{pmatrix} a_2 & b_2 \\ 0 & d_2 \end{pmatrix},$$

with $p_1 \log |a_1| + p_2 \log |a_2| > 0$ and $b_j \neq 0$, for some $j = 1, 2$. The inverse A_λ^{-1} of the cocycle introduced in Example 2.2 provides a particular case of the example discussed above.

4. *Not quasi-irreducible:* Consider the diagonal cocycle $A \in SL_2(\mathbb{R})$ given by

$$A_1 = \begin{pmatrix} \beta & 0 \\ 0 & \beta^{-1} \end{pmatrix} \quad \text{and} \quad A_2 = \begin{pmatrix} \beta^{-1} & 0 \\ 0 & \beta \end{pmatrix},$$

with $\beta > 1$. Notice that $L(A) > 0$ if and only if $p_1 \neq p_2$.

Remark 2.2 We mentioned before the relation between the Lyapunov exponent of Schrödinger cocycles and the spectrum of the respective Schrödinger operator. In the Anderson model presented above, the spectrum of Schrödinger operator is completely determined and is given by the set

$$\text{Spec}(H_x) = [a_1 - 2, a_1 + 2] \cup [a_2 - 2, a_2 + 2].$$

See [28] for more details.

Now we are ready to collect properties of the stationary measures for most $SL_2(\mathbb{R})$ cocycles which are described in some of the items of the next proposition. In what follows, for every matrix $B \in SL_2(\mathbb{R})$, we denote by B^t its transpose matrix.

Proposition 2.2 *Let $A = (A_1, \ldots, A_\kappa) \in SL_2(\mathbb{R})^\kappa$ be a locally constant cocycle.*

1. *The set of strongly irreducible cocycles is a countable intersection of open and dense subsets of $SL_2(\mathbb{R})^\kappa$;*
2. *The cocycle A is strongly irreducible if and only if all the stationary measures are non atomic;*
3. *(Furstenberg's criterion) If the semigroup generated by $A_1, \ldots, A_\kappa$ is unbounded and A is strongly irreducible, then $L(A) > 0$;*
4. *Assume that A is irreducible and $L(A) > 0$, then A is strongly irreducible;*
5. *If $L(A) > 0$ and A is quasi-irreducible, then there exists a unique stationary measure.*
6. *If $L(A) > 0$, then for μ-a.e. $x \in \Sigma$,*

$$A^n(x)^t_*(Leb) = \left(A(x)^t \cdots A(\sigma^{n-1}(x))^t\right)_* (Leb) \xrightarrow{*} \delta_{e^-(x)^\perp}.$$

Moreover, there exists at most one non-atomic stationary measure. If η is a non-atomic stationary measure, then for μ-a.e. $x \in \Sigma$,

$$A^n(x)^t_* \eta \xrightarrow{*} \delta_{e^-(x)^\perp}.$$

7. *If $L(A) > 0$ and A is not irreducible, then either A (or A^{-1}) is quasi-irreducible or A is conjugated to a diagonal cocycle.*
8. *If $L(A) > 0$, then there exists at most two ergodic stationary measures.*
9. *Assume that A is strongly irreducible and let η be the stationary measure. Then $L(A) > 0$ if and only if there exists $y \in \Sigma$ and $\hat{v} \in \mathbb{P}^1$ such that*

$$A^n(y)_*(Leb) \xrightarrow{*} \delta_{\hat{v}}.$$

In this case η is unique and there exists a measurable map $\xi : \Sigma \to \mathbb{P}^1$ such that for μ-a.e. $x \in \Sigma$

$$\eta = \int_\Sigma \delta_\xi \, d\mu.$$

We also have that the set supp η *is the unique minimal set for the action of the semi-group generated by $(A_1, \ldots, A_\kappa)$.*

Outline of the Proof and References Item 1 is a consequence of the fact that the set

$$\mathcal{V}_n = \left\{ A \in SL_2(\mathbb{R})^\kappa : \nexists \hat{V} \subset \mathbb{P}^1,\ |\hat{V}| \leq n \text{ with } A_i(\hat{V}) = \hat{V}, \text{ for all } 1 \leq i \leq \kappa \right\},$$

is open and dense in $SL_2(\mathbb{R})$. The union of $\mathcal{V}_n$ for all $n \geq 1$ is the set of strong irreducible cocycle. Notice in particular that the set $\mathcal{V}_1$, which is the set of irreducible cocycles, is open and dense in $SL_2(\mathbb{R})^\kappa$.

For item 2, notice that if A preserves a finite set, say $\hat{V}$, with minimal cardinality, then we can build the atomic stationary measure $\eta := \sum_{\hat{v} \in \hat{V}} 1/|\hat{V}| \cdot \delta_{\hat{v}}$. For the other implication, see [69, Lemma 6.9]. Item 3 can be found in [69, Theorem 6.11]. For a proof of 4, see [22, Theorem 6.1]. Item 5 can be found in [30, Proposition 4.2]. For a proof of item 6 consult [69, Section 6.3.2] and [37, Section 1.8].

To see item 7, assume that A is not irreducible and $L(A) > 0$. Then A preserves a direction $\hat{v} \in \mathbb{P}^1$ and so $\delta_{\hat{v}}$ is an ergodic stationary measure for A. If $L(A|_{\hat{v}}) = L(A)$, then either A is quasi-irreducible or admits another invariant direction and so in this case is conjugated to a diagonal cocycle. Therefore, we may assume that $L(A|_{\hat{v}}) = -L(A)$. Let η be an ergodic stationary measure for A, with,

$$L(A) = \int_{\Sigma \times \mathbb{P}^1} \log \|A(x)\, v\| \, d(\mu \times \eta)(x, \hat{v}),$$

and $\eta \neq \delta_{\hat{v}}$. Consider the quantity $t = \max\{\eta(\hat{v})\colon \hat{v} \in \mathbb{P}^1\}$, and define $\hat{V}_t = \{\hat{v} \in \mathbb{P}^1 : \eta(\hat{v}) = t\}$. Using that η is stationary is easy to see that $\hat{V}_t$ is preserved by A_i, for every $i = 1, \ldots, \kappa$. Since $L(A) > 0$, we see that there exists a hyperbolic matrix in the semi-group generated by $A_1, \ldots, A_\kappa$. This imposes a restriction on the number of elements of $\hat{V}_t$, i.e., $|\hat{V}_t| \leq 2$ (one of the invariant directions of the hyperbolic matrix is $\hat{v}$). Notice that $\hat{v} \notin \hat{V}_t$ since otherwise, by ergodicity of η, we would have $\eta = \delta_{\hat{v}}$. Then, $|\hat{V}_t| = 1$ and so A is conjugated to a diagonal. The conclusion, therefore is that if A is not conjugated to a diagonal, then η is non-atomic. But, using item 6 we see that η is unique with such property. In particular, $\mathrm{Stat}(A) = [\delta_{\hat{v}}, \eta]$ and A^{-1} is quasi-irreducible.

Item 8 may be obtained from 5, the proof of item 7 and the observation that for diagonal cocycles there are only two ergodic stationary measures. Item 9 is due to Guivarc'h and Raugi [43] and can be recovered as combinations of a few results in [37] (see Theorem 1.23, Lemma 1.30 and Theorem 1.34 in [37] and references therein). □

Example 2.4 It is easy to see that the set of strongly irreducible cocycles is not open. Indeed, let $\theta \in \mathbb{R}\backslash\mathbb{Q}$ and consider p_k/q_k sequence of rationals converging to θ. Then the sequence of cocycles $A_{1,k} = R_{2\pi p_k/q_k}$ are not strongly irreducible and it converges to the cocycle $A_1 = R_{2\pi\theta}$ which is strongly irreducible.

Lets summarize the content of the above proposition. Assume that $L(A) > 0$. If A is irreducible, then A is strongly irreducible and thus there exists a unique stationary measure (see item 1 of Example 2.3).

In other direction, if A is *reducible*, i.e., there exists an invariant direction, then A is conjugated to a triangular cocycle. So, we may assume that $A \in SL_2(\mathbb{R})^\kappa$ is given by matrices, $A_i = \begin{pmatrix} a_i & b_i \\ 0 & d_i \end{pmatrix}$, for every $i = 1, \ldots, \kappa$. The measure $\delta_{\hat{e}_1}$ is clearly a stationary measure for $\hat{A}$. Assume first that $\delta_{\hat{e}_1}$ is the only stationary measure. So, by Furstenberg's formula, $L(A|_{\hat{e}_1}) = L(A)$ and in particular A is quasi-irreducible. Now, if there are two ergodic stationary measures, then we have the following situations: either the other ergodic stationary measure is atomic and so A is a diagonal cocycle, or there is a non-atomic ergodic stationary measure η satisfying

$$\int_{\Sigma\times\mathbb{P}^1} \log \|A(x)\, v\|\; d(\mu \times \eta)(x, \hat{v}) = L(A), \tag{5}$$

and $L(A|_{\hat{e}_1}) = -L(A)$ and so A^{-1} is quasi-irreducible (that is the case, for instance, of the Example 2.2).

Example 2.5 All quasi-irreducible uniformly hyperbolic cocycles have a unique forward and a unique backward stationary measure. These measures have disjoint support on $\mathbb{P}^1$ and are not necessarily singular with respect to Lebesgue (see Example 2.2). In [3] we have an explicit description of the *multicones* containing the support of these measures.

The situation for zero Lyapunov exponent is trickier. For example, we could consider the cocycle given by the matrices $A_i = I$ for every $i = 1, \ldots, \kappa$, where $I \in SL_2(\mathbb{R})$ is the identity matrix. Here, every measure on $\mathbb{P}^1$ is stationary.

On the other hand, we have the situation in which the stationary measures are unique. That is the case for example when A is strongly irreducible and the semi-group generated by $A_1, \ldots, A_\kappa$ is bounded. Then $L(A) = 0$ and there is only one stationary measure. This measure is indeed equivalent to the Lebesgue measure on $\mathbb{P}^1$.

Another interesting example of a cocycle with zero Lyapunov exponent is given by Kifer's example (Example 2.1). This cocycle is unbounded and irreducible, but it is not strongly irreducible (the set of directions $\{\hat{e}_1, \hat{e}_2\}$ is preserved by the action of A). The unique stationary measure is given by the measure $\eta = 1/2 \cdot \delta_{\hat{e}_1} + 1/2 \cdot \delta_{\hat{e}_2}$.

Cocycles with zero Lyapunov exponent present some type of rigidity. That is the content of the so called *Invariance Principle:*

Theorem 2.2 (Ledrappier [51]) *Let $A = (A_1, \ldots, A_\kappa)$ be a locally constant cocycle. If $L(A) = 0$ and η is a stationary measure for A, then $(A_i)_*\eta = \eta$, for every $i = 1 \ldots, \kappa$.*

One approach to obtain such a result is through the notion of *Furstenberg's entropy* of a given stationary measure η which for a cocycle A is defined by the quantity

$$h_A(\eta) := \sum_{i=1}^{k} p_i \int_{\mathbb{P}^1} -\log \frac{d(A_i^{-1})_*\eta}{d\eta} d\eta. \tag{6}$$

This quantity is a measurement of the lack of invariance of the stationary measure η by the A_i-action: $h_A(\eta) = 0$ if and only if $(A_i)_*\eta = \eta$ for every $i = 1, \ldots, \kappa$. Theorem 2.2 is therefore a consequence of the following inequality

$$0 \leq h_A(\eta) \leq 2L(A).$$

See [69, Theorem 7.2] for more details about this proof.

Remark 2.3 The rigidity presented in Theorem 2.2 is not a sufficient condition to guarantee zero Lyapunov exponent. Indeed, if $A = (A_1, \ldots, A_\kappa) \in SL_2(\mathbb{R})^\kappa$ is a diagonal cocycle, meaning that $A_i A_j = A_j A_i$ for all $i, j = 1 \ldots, \kappa$, then all the stationary measures are preserved by all the matrices A_i, but we could have positive Lyapunov exponent (see for item 4 of Example 2.3).

We can now say a bit more about the locally constant cocycles with zero Lyapunov exponent.

Proposition 2.4 *Let $A \in SL_2(\mathbb{R})^\kappa$ and assume that $L(A) = 0$.*

1. *If A is strongly irreducible, then there exists a unique stationary measure. This measure is absolutely continuous with respect to Lebesgue;*

2. *If A is irreducible and the semi-group generated by A is unbounded, then there is a unique stationary measure $\eta = 1/2\delta_{\hat{u}} + 1/2\delta_{\hat{v}}$ for some $\hat{u}, \hat{v} \in \mathbb{P}^1$. Moreover, up to conjugation, A_i is either diagonal or a rotation by $\pi/2$ and both cases must to occur;*
3. *If A is reducible, then, up to conjugation, $A_i = \begin{pmatrix} a_i & b_i \\ 0 & d_i \end{pmatrix}$ for every $i = 1, \ldots, \kappa$ and*

$$a_1^{p_1} \cdots a_\kappa^{p_k} = 1.$$

If $b_1 = \cdots = b_\kappa = 0$, i.e., A_i is diagonal for every i and not always equal to identity, then we have two ergodic stationary measures, namely $\eta_1 = \delta_{\hat{e}_1}$ and $\eta_2 = \delta_{\hat{e}_2}$ and so Stat$(A) = [\eta_1, \eta_2]$. If, otherwise, there exists $b_i \neq 0$, then $\eta = \delta_{\hat{e}_1}$ is the unique stationary measure.

Outline of the Proof If A is strongly irreducible, by Furstenberg's criteria we have that the group G generated by $A_1, \ldots, A_\kappa$ is contained in a compact subgroup of $SL_2(\mathbb{R})$. This implies that, up to change of the norm in $\mathbb{R}^2$, we may assume that the matrices in G are orthogonal. In particular, Lebesgue measure on $\mathbb{P}^1$ is the unique stationary measure. This finishes item 1.

For item 2, observe that the semi-group being unbounded implies that there exists a sequence of elements B_n in the semi-group such that $\|B_n\| \to \infty$. By Theorem 2.2, $(B_n)_*\eta = \eta$. So, up to taking a subsequence, the projective action of B_n converges to a quasi projective map P (see [20]) that has a one dimensional kernel $\hat{v}$ and one dimensional image $\hat{u}$. Thus η must have the form $a\,\delta_{\hat{v}} + b\,\delta_{\hat{u}}$. This implies that $\{\hat{v}, \hat{u}\}$ is invariant.

As A is irreducible, there exists some matrix A_i that exchanges $\hat{u}$ and $\hat{v}$, so $a = b = 1/2$. Up to changing the canonical basis to u and v, A_i are diagonal or exchange both (rotation of $\pi/2$). Thus, one of the matrices A_i is a rotation and since the cocycle A is unbounded, another matrix A_j must to be hyperbolic and diagonal.

For item 3, the first part is a direct consequence of the existence of an invariant direction and the fact that the exponent is equal to $\sum p_i \log |a_i|$. The second part is consequence of the invariance of η for every element of the group and the fact that for a parabolic matrix there exists a unique invariant measure. □

Remark 2.4 In the case that A is irreducible, A can have infinitely many stationary measures. That is the case, for example, for the groups generated by a single matrix $A = R_{2\pi\theta}$, with $\theta \in \mathbb{Q}\backslash(\mathbb{Z} \cup \mathbb{Z}/2)$.

2.3 Regularity for Locally Constant

We have seen a few results with criteria for the positivity of the Lyapunov exponent (Problem 2.1) in the context of locally constant cocycles. Now, we turn our attention to the problem of regularity of the function L.

We start with the simplest question: is the function that associates each $A \in SL_2(\mathbb{R})^\kappa$ its Lyapunov exponent $L(A)$ continuous?

The approach to handle this question goes as follows: let $(A_n)_n \subset SL_2(\mathbb{R})^\kappa$ be a sequence of locally constant cocycles converging to $A \in SL_2(\mathbb{R})^\kappa$. Since the points with zero Lyapunov exponents are continuity points of L, we may assume that $L(A) > 0$.

For each $k \geq 1$, consider an (ergodic) stationary measure η_k for A_k satisfying that

$$\int_{\Sigma\times\mathbb{P}^1} \log \|A_k(x)\, v\|\; d(\mu \times \eta_k)(x, \hat{v}) = L(A_k). \tag{7}$$

Up to taking a subsequence, the limit of η_k in the weak* topology exists and is a stationary measure η for the limit cocycle A (not necessarily ergodic). The problem is then reduced to prove that

$$\int_{\Sigma\times\mathbb{P}^1} \log \|A(x)\, v\|\; d(\mu \times \eta)(x, \hat{v}) = L(A). \tag{8}$$

Using the above strategy and the uniqueness of the stationary measure when A is quasi-irreducible we have the following result.

Theorem 2.3 (Furstenberg and Kifer [39]) *If A (or A^{-1}) is quasi-irreducible, then A is a continuity point of L.*

We proceed with the analysis of continuity of the Lyapunov exponent of cocycles A such that neither A or A^{-1} are quasi-irreducible. Then, up to a change of coordinates, we may assume that A is diagonal, i.e., $A_i = \begin{pmatrix} a_i & 0 \\ 0 & d_i \end{pmatrix}$ for every $i = 1, \ldots, \kappa$. Hence, in this case, $\mathrm{Stat}(A) = [\delta_{\hat{e}_1}, \delta_{\hat{e}_2}]$ with, say, $L(A|_{\hat{e}_1}) = L(A) = -L(A|_{\hat{e}_2})$. Write

$$\eta = q_1\delta_{\hat{e}_1} + q_2\delta_{\hat{e}_2}. \tag{9}$$

If $q_2 = 0$, then equality (8) is satisfied. So, we just need to deal with the case $q_2 \neq 0$. The fact that $L(A|_{\hat{e}_2}) = -L(A) < 0$ implies that $\hat{e}_2$ is a *p-expanding* fixed point for the projective action of A, i.e, it expands on average around $\hat{e}_2$. Furthermore, this behavior passes to nearby cocycles (although $\hat{e}_2$ will not be fixed by them) A_k, for k large. This indicates that η_k concentrates mass around $\hat{e}_2$, so we expect that η_k has an atom close to $\hat{e}_2$. That is exactly the case, and the formalization is provided by the so called *energy argument*: if η_k is non-atomic for a subsequence of k's and $\hat{e}_2$ is p-expanding, then we should have $\eta(\{\hat{e}_2\}) = 0$ (see [69, Section 10.4]).

So, we may assume that η_k has an atom for every k. In particular, A_k is not strongly irreducible (see Item 2 of Proposition 2.2). Let $\hat{V}_k$ be a finite set of projective directions which is invariant by the coordinates of A_k. The fact that $L(A) > 0$ implies that there exists $z \in \Sigma$ and $m \in \mathbb{N}$ such that $A^m(z)$ is a hyperbolic matrix. Then $A_k^m(z)$ for every k (sufficiently large) is also hyperbolic. This guarantees that $|\hat{V}_k| \leq 2$. Assume that for every $k \geq 1$, $\hat{V}_k = \{\hat{v}_k\}$. Up to taking a subsequence, $\hat{v}_k$ converges to $\hat{v} \in \{\hat{e}_1, \hat{e}_2\}$.

Using that $\delta_{\hat{v}_k} = \eta_k \to \eta = \delta_{\hat{v}}$ and Eq. (9), jointly with the fact that $q_2 \neq 0$, we see that we must to have $q_1 = 0$, i.e., $\hat{v} = \hat{e}_2$. Hence $L(A|_{\hat{v}}) = -L(A)$. But, since $\eta_k = \delta_{\hat{v}_k}$ for every k, by Eq. (7) we have that $L(A_k|_{\hat{v}_k}) \geq 0$. This contradiction implies that $q_2 = 0$. The case where there are infinitely many k such that $|\hat{V}_k| = 2$ is handled similarly.

The next result due to Bocker and Viana summarizes the above discussion.

Theorem 2.4 (Bocker and Viana [18]) *The function $L : SL_2(\mathbb{R})^\kappa \to \mathbb{R}$ is continuous.*

Surprisingly, when compared with the general Problem 2.1 for continuous cocycles (see Theorem 1.2), in the world of locally constant we always have continuity of the Lyapunov exponent. That finishes the soft analysis of the function L.

Now we discuss the hard analysis version of the Problem 2.1 for locally constant cocycles. In other words, we study the modulus of continuity of the Lyapunov exponent. The first result to be mentioned in this direction is due E. Le Page.

Theorem 2.5 (Le Page [56]) *Let $I \subset \mathbb{R}$ be a compact interval and let $\lambda \mapsto A_\lambda \in SL_2(\mathbb{R})^\kappa$ be a γ-Hölder continuous one-parameter family of locally constant cocycles, $\gamma > 0$. Assume that for every $\lambda \in I$, A_λ is strongly irreducible and generates an unbounded semi-group. Then, the map*

$$I \ni \lambda \mapsto L(A_\lambda)$$

is locally Hölder continuous.

Remark 2.5 The Hölder exponent of the conclusion may be smaller than the γ in the assumption.

One reason for the choice of a one parameter family in the previous result is the immediate application of the theorem to guarantee continuity of the Lyapunov exponent with respect to the energy in the Anderson model (see item 1 in the Example 2.3 above). This result was later generalized by Duarte and Klein, where a machinery to obtain a modulus of continuity of the Lyapunov exponent was developed.

Theorem 2.6 (Duarte and Klein [32]) *Let $A \in SL_2(\mathbb{R})^\kappa$ be a quasi-irreducible cocycle with positive Lyapunov exponent. Then, there exist $\theta > 0$ and a neighborhood $\mathcal{U}(A)$ of A in $SL_2(\mathbb{R})^\kappa$ such that $L : \mathcal{U}(A) \to \mathbb{R}$ is θ-Hölder continuous.*

Now we discuss the strategy to obtain Theorem 2.6. The proof contains three main steps:

Step 1: Analysis of spectral properties of the Markov operator $Q_A : C^0(\mathbb{P}^1) \to C^0(\mathbb{P}^1)$,

$$Q_A(\varphi)(\hat{v}) = \sum_{i=1}^{\kappa} p_i\, \varphi(A_i \hat{v}).$$

Let us elaborate a bit more on that. We consider for each $\theta \in (0, 1)$ the space of θ-Hölder continuous functions $C^\theta(\mathbb{P}^1)$ with the norm

$$\|\varphi\|_\theta := \|\varphi\|_\infty + [\varphi]_\theta \quad \text{where} \quad [\varphi]_\theta := \sup_{\hat{v} \neq \hat{u}} \frac{|\varphi(\hat{v}) - \varphi(\hat{u})|}{d(\hat{v}, \hat{u})^\theta}.$$

The general idea is to prove that there exists $\theta \in (0, 1)$ such that the operator Q_A preserves $C^\theta(\mathbb{P}^1)$ and when restricted to this space is *quasi-compact*. In other words, denoting by η the (unique) stationary measure for A and

$$F_\eta = \left\{ \varphi \in C^\theta(\mathbb{P}^1) \colon \int_{\mathbb{P}^1} \varphi\, d\eta = 0 \right\}.$$

There exists a number $\rho \in (0, 1)$ such that

$$\operatorname{Spec}(Q_A) = \{1\} \cup \operatorname{Spec}(Q_A|_{F_\eta}) \quad \text{and} \quad |\operatorname{Spec}(Q_A|_{F_\eta})| < \rho < 1$$

(the eigenvalue 1 is associated with the constant functions). That is a consequence of the fact that for n sufficiently large Q_A^n contracts the θ-Hölder semi-norm $[\cdot]_\theta$. To see that, we observe that for every $\varphi \in C^\theta(\mathbb{P}^1)$,

$$[Q_A^n(\varphi)]_\theta \le [\varphi]_\theta \sup_{\hat{u} \neq \hat{v}} \int_\Sigma \left(\frac{d(A^n(x)\,\hat{u},\, A^n(x)\,\hat{u})}{d(\hat{u}, \hat{v})} \right)^\theta d\mu(x).$$

So, the contraction property will be a consequence of the (exponential) decay of the following quantities:

$$\sup_{\hat{u} \neq \hat{v}} \int_\Sigma \left(\frac{d(A^n(x)\,\hat{u},\, A^n(x)\,\hat{u})}{d(\hat{u}, \hat{v})} \right)^\theta d\mu(x) = \sup_{\hat{v} \in \mathbb{P}^1} \int_\Sigma \frac{1}{\|A^n(x)\, v\|^{2\theta}}\, d\mu(x). \tag{10}$$

It is at this point that the assumptions that $L(A) > 0$ and A is quasi-irreducible are used. These guarantee that the limit

$$\lim_{n \to \infty} \frac{1}{n} \int_\Sigma \log \left\| A^n(x)\, v \right\| \, d\mu(x) = L(A), \tag{11}$$

is *uniform* in the unitary vector $v \in \mathbb{R}^2$ (compare with Oseledet's Theorem). This interesting fact comes from a more general result due to Furstenberg and Kifer in [39] which is a non-random version of Oseledet's Theorem (see also [50]). That is the main tool to ensure the exponential decay in n of the quantity in (10).

Remark 2.6 This type of decay was used in Y. Perez result [58]. See the discussion just after the statement of Theorem 2.1.

Step 2: Establish uniform large deviation estimates: once we have the quasi-compactness operator we may use standard techniques of additive random process to prove the following type of estimate: there exist constants $\delta,\ C,\ \kappa,\ \varepsilon_0 > 0$ such that for every cocycle $B \in SL_2(\mathbb{R})^\kappa$ with $\|A - B\| < \delta$, for every $\varepsilon \in (0, \varepsilon_0)$ and for every $n \in \mathbb{N}$

$$\mu\left(\left\{x \in \Sigma\colon \left|\frac{1}{n}\log\|B^n(x)\| - L(B)\right| > \varepsilon\right\}\right) \leq C\, e^{-\kappa\varepsilon^2 n}.$$

Step 3: Combine uniform large deviation estimates with accurate analysis of the geometry of the projective action of "very hyperbolic" matrices $A^n(x)$. The idea here is to use a process of exclusion of sequences $x, y \in \Sigma$ such that $B^n(x)$ and $B^n(y)$ are very hyperbolic (exponentially large norm) but the product $B^n(x)\, B^n(y)$ is small. Using a tool called *avalanche principle* (see [30]) combined with the uniform large deviation estimate it is possible to show that this process of exclusion of sequences only eliminates a small probability subset of Σ. This is enough to have very good control of the finitary differences $|\log\|B^n(x)\| - \log\|A^n(x)\||$ for suitable scales n and most of the sequences $x \in \Sigma$. This provides the desired modulus of continuity for the Lyapunov exponent function.

The next example indicates that, even though, under the assumption of $L(A) > 0$ and A strongly irreducible, we have Hölder regularity in a neighborhood of A, the optimal Hölder exponent can get arbitrarily close to zero.

Example 2.6 Simon and Taylor [63]] Here, we fix $\kappa = 2$. Consider real numbers $0 < a_0 < a_1$ and define

$$A_0 = \begin{pmatrix} a_0 & -1 \\ 1 & 0 \end{pmatrix} \quad \text{and} \quad A_1 = \begin{pmatrix} a_1 & -1 \\ 1 & 0 \end{pmatrix}.$$

Let A be the locally constant cocycle generated from A_0 and A_1. A is unbounded and strongly irreducible. So, by Furstenberg's criteria $L(A) > 0$. Thus, applying Theorem 2.6, we conclude that there exists $\theta \in (0, 1)$ such that L is θ-Hölder continuous in a neighborhood of A. However, Halperin/Simon and Taylor showed that if

$$\theta_0 > \frac{2 \log 2}{\cosh^{-1}\left(1 + \frac{a_1 - a_0}{2}\right)},$$

then L is not θ_0-Hölder continuous. In particular, making $|a_1 - a_2| \to \infty$ we can build examples of cocycles such that the Hölder exponent converges to 0.

The case of zero Lyapunov exponent is different. It is possible that the Lyapunov exponent is not even Hölder continuous.

Example 2.7 (Duarte et al. [34]) Take A given by Kifer's example (Example 2.1) with $p = (1/2, 1/2)$ and $\alpha = 2$. This is an example with zero Lyapunov exponent. Using Halperin/Simon and Taylor's strategy Duarte, Klein and Santos showed that the Lyapunov exponent is not even β-Hölder continuous at A for any $\beta > 0$. Actually, they proved that the best regularity that we could expect is a weak version of Hölder regularity called log-Hölder regularity (see conclusion of item 2 of Theorem 2.7).

An improvement in understanding the behavior of the Lyapunov exponent for locally constant cocycles was provided by E. Tall and M. Viana as described in the next result. Notice that there is no irreducibility assumption of any kind.

Theorem 2.7 (Tall and Viana [66]) *It holds that,*

1. *(Pointwise Hölder) Assume that $L(A) > 0$. Then, there exist a neighborhood $\mathcal{U}(A) \subset SL_2(\mathbb{R})^\kappa$ of A, $C > 0$ and $\theta > 0$ such that for every $B \in \mathcal{U}(A)$,*

$$|L(A) - L(B)| \leq C \, \|A - B\|^\theta .$$

2. *(Pointwise log-Hölder) For every $A \in SL_2(\mathbb{R})^\kappa$, there exist a neighborhood $\mathcal{U}(A) \subset SL_2(\mathbb{R})^\kappa$ of A, $C > 0$ and $\theta > 0$ such that for every $B \in \mathcal{U}(A)$,*

$$|L(A) - L(B)| \leq C \log\left(\frac{1}{\|A - B\|}\right)^{-\theta} .$$

The proof of this result is a consequence of a careful analysis of the phenomena presented in the discussion just before Theorem 2.4. The techniques are based on many classical probabilistic results such as the central limit theorem and the diffusion power law.

Due to the technical level of the proof of Theorem 2.7, we do not discuss its details here. Instead, we provide a result of similar flavour which is the pointwise Lipschitz continuity of the Lyapunov exponent at a strongly irreducible cocycle $A \in SL_2(\mathbb{R})^\kappa$ with $L(A) = 0$ (the argument can be found in [34] or in [66]). Indeed, by Furstenberg's criterion, the closure G of the group generated by the coordinates of A is a compact subgroup of $SL_2(\mathbb{R})$. So, there exists a norm on $\mathbb{R}^2$ such that the matrices on G are orthogonal. Let $\|\cdot\|'$ be the associated operator norm. Notice that for any cocycle $B \in SL_2(\mathbb{R})^\kappa$,

$$0 \leq L(B) = \lim_{n\to\infty} \frac{1}{n} \int_\Sigma \log \|B^n\|' \, d\mu \leq \int_\Sigma \log \|B\|' \, d\mu \text{ and } \int_\Sigma \log \|A\|' \, d\mu = 0.$$

Then,

$$|L(B) - L(A)| = L(B) \leq \int_\Sigma \log \|B\|' \, d\mu \leq \int_\Sigma \log\left(1 + \|B - A\|'\right) d\mu$$

$$\leq \int_\Sigma \|B - A\|' \, d\mu \leq C \, \|A - B\| \, .$$

The non-pointwise continuity was treated by Duarte and Klein as described by the next result.

Theorem 2.8 (Duarte and Klein [31]) *Assume that $L(A) > 0$. Then, there exists a neighborhood $\mathcal{U}(A) \subset SL_2(\mathbb{R})$ of A such that $L : \mathcal{U}(A) \to \mathbb{R}$ is weak-Hölder continuous. More precisely, there exist constants $C, \alpha, \beta > 0$ such that for every $B_1, B_2 \in \mathcal{U}(A)$ we have*

$$|L(B_1) - L(B_2)| \leq C \exp\left(-\alpha \left(\log \frac{1}{\|B_1 - B_2\|}\right)^\beta\right).$$

The proof of this result follows the same general strategy as described in the proof of Theorem 2.6. The idea is to establish some version of uniform large deviation estimates in a neighborhood of A. The result only provides a weak-Hölder regularity. This is a consequence of the fact that large deviation estimates obtained in Theorem 2.8 are no longer of exponential type but only sub-exponential. These worse estimates are due the lack of uniformity of the convergence in the limit (11) in the diagonal case (the only case that still needs analysis).

The idea to proof Theorem 2.8 goes as follows. If A is diagonal with $L(A) > 0$, then we have classical large deviation estimates (LDE) to A which is uniform in a neighborhood of A among the diagonal cocycles. If $B \in SL_2(\mathbb{R})^\kappa$ is a quasi-irreducible cocycle near A, then until certain scale n_1 the finite scale Lyapunov exponent, $\frac{1}{n_1} \log \|B^{n_1}\|$, absorb the property from the diagonal and satisfy LDE. By the quasi-irreducibility of B, this cocycle also satisfy a LDE, but that could only be seen after a different scale n_2 with possibly $n_2 \gg n_1$. From here, the main idea is to observe that this scales are not that far from each other. Indeed, $n_2 - n_1 \leq O(\log \delta^{-1})$, where δ is a distance from B to the diagonal cocycles. This is enough to save the LDE between scales n_1 and n_2 (bridge argument) although the exponential type could be lost in the process.

2.4 Sharp Modulus of Continuity

As we could see earlier, with the exception of uniformly hyperbolic cocycles, the Lyapunov exponent can have a rather bad modulus of continuity. For irreducible $A \in SL_2(\mathbb{R})^\kappa$ with positive Lyapunov exponent (open and dense set of cocycles) we have seen (Theorem 2.6) that L is θ-Hölder continuous in a neighborhood of A. In this subsection, we discuss geometric obstructions that provides bounds for how large the Hölder exponent θ can be.

Let $A = (A_1, \dots, A_\kappa) \in SL_2(\mathbb{R})^\kappa$ and denote by $\langle A\rangle^+$ the semi-group generated by the matrices $A_1, \dots, A_\kappa$. We say that the cocycle A admits a *heteroclinic tangency* if there exist matrices $B, T, D \in \langle A\rangle^+$ such that B, D are hyperbolic and $T\,\hat{v}_+(B) = \hat{v}_-(D)$, where $\hat{v}_+(B), \hat{v}_-(D)$ denotes the eigen-directions associated respectively to the largest eigenvalue of B and the smallest eigenvalue of D. In this case, we say that the triple (B, T, D) is a heteroclinic tangency for the cocycle A.

Example 2.8 (Heteroclinic Tangecies) Consider the following examples.

1. Let A be the Kifer's example (Example 2.1). Then, the triple, (A_1, A_2, A_1) is a tangency for A;
2. Let $A_{\alpha,\beta}$ be the cocycle in the Example 3.1. Then, the triple (A_1, A_1, A_2) is a heteroclinic tangency for $A_{\alpha,\beta}$.
3. Fix $\kappa = 3$ and consider the cocycle $A \in SL_2(\mathbb{R})^3$ given by the matrices

$$A_1 = \begin{pmatrix} \beta & 0 \\ 0 & \beta^{-1} \end{pmatrix}, \quad A_2 = R_{2\pi\theta}\, A_1\, R_{-2\pi\theta} \quad \text{and} \quad A_3 = R_{\pi(4\theta+1)/2}.$$

where $\theta \in (0, 1/2)$ with $\theta \neq 1/4$. Then A is strongly irreducible and the triple (A_1, A_3, A_2) a tangency for A.

Proposition 2.12 *It holds that,*

1. *If A is uniformly hyperbolic, then A does not admits a heteroclinic tangency;*
2. *If (B, T, D) is a tangency for A, then (B^n, T, D^m) is also a tangency for A;*
3. *If $A \in SL_2(\mathbb{R})^\kappa$ is the boundary of the uniformly hyperbolic cocycles and $\langle A\rangle^+$ does not contain a parabolic matrix, then A admits a heteroclinic tangency;*
4. *Cocycles $A \in SL_2(\mathbb{R})^\kappa$ admitting a heteroclinic tangency are dense outside the set of the uniformly hyperbolic cocycles.*
5. *If (B, T, B) is a tangency for A, then $\langle A\rangle^+$ contains an elliptic element.*

Outline of the Proof and References Item 1 is a consequence of the multicone characterization of the uniformly hyperbolic locally constant cocycles in [3, Theorem 2.2]. Item 2 is a direct consequence of the definition. Item 3 can be found in [3, Theorem 4.1]. For a proof of item 4 see [14, Section 7]. For item 5 see [3, Remark 4.2]. □

The existence of tangencies will be the main tool to design perturbations that cause a drastic drop in the value of the Lyapunov exponent. These perturbations are carried out within carefully chosen one parameter families of cocycles $\{A_t\}_{t\in I} \subset SL_2(\mathbb{R})^\kappa$ where the Lyapunov exponent variation can be measured in terms of the so called *fibered rotation number* of the family.

More precisely, we say that a family of locally constant cocycles $\{A_t\}_{t\in I} \subset SL_2(\mathbb{R})^\kappa$ is (strictly) *positively winding* (or *monotone*) if there exists $c_0 > 0$ and $n_0 \in \mathbb{N}$ such that for every $n > n_0$, $x \in \Sigma$, $\hat{v} \in \mathbb{P}^1$ and every $t \in I$,

$$\frac{d}{dt} A_t^n(x)\,\hat{v} \geq c_0 > 0 > -c_0 \geq \frac{d}{dt} A_t^{-n}(x)\,\hat{v}.$$

For such family and for any subinterval $J \subset I$, the limit

$$\rho(J) = \lim_{n\to\infty} \frac{1}{n\pi} \ell_J(A_t^n(x)\,\hat{v}), \tag{12}$$

exists and is constant for μ-a.e. $x \in \Sigma$ and $\hat{v} \in \mathbb{P}^1$. Here, $\ell_J(A_t^n(x)\,\hat{v})$ denotes the length of the projective curve $J \ni t \mapsto A_t^n(x)\,\hat{v}$. The expression (12) above defines a measure ρ called *fibered rotation measure* of the family $\{A_t\}_{t\in I} \subset SL_2(\mathbb{R})^\kappa$ (for more details see [41], [42] or [13] and references therein).

Example 2.9 Consider the family of cocycles $\{A_E\}_{E\in\mathbb{R}} \subset SL_2(\mathbb{R})^\kappa$ provided by the Anderson model, i.e., for each $i = 1, \ldots, \kappa$, $A_{i,E} = \begin{pmatrix} a_i - E & -1 \\ 1 & 0 \end{pmatrix}$. By a direct computation, this family is positively winding (with $n_0 = 2$). In this case, up to a normalization constant, the fibered rotation number coincides with the integrated density of states which is the distribution measure of the spectrum of the Schrodinger operators H_x (see Example 1.3).

In what follows, $H(\mu) := \sum_{i=1}^{\kappa} -p_i \log p_i$ denotes the *Shannon entropy* associated to the measure μ.

Theorem 2.9 ([14] and [13]) *Let $\{A_t\}_{t\in I} \subset SL_2(\mathbb{R})^\kappa$ be a positive winding family of locally constant cocycles. Assume that for some $t_0 \in I$,*

1. *A_{t_0} is strongly irreducible and $\langle A\rangle^+$ is unbounded;*
2. *A_{t_0} admits a heteroclinic tangency.*

Then, ρ is not θ-Hölder continuous for any $\theta > H(\mu)/L(A_{t_0})$.

We stress here that, by Theorem 2.6, the above assumptions guarantee that L is θ-Hölder continuous in a neighborhood of A_{t_0} in $SL_2(\mathbb{R})^\kappa$ for some $\theta \in (0, 1)$. The above result provides a limitation for such θ when $H(\mu)/L(A_{t_0}) < 1$.

Remark 2.7 In [14], Theorem 2.9 is proved, but in the context of families Schrödinger cocycles and the integrated density of states (see Example 2.9). After careful adaptation of the techniques in [14] the more general version of the result, Theorem 2.9, was obtained in [13].

Before we discuss the ideas behind the proof of Theorem 2.9, we mention how to build families of cocycles satisfying the assumptions of this theorem and how to use these families to have a similar result for the Lyapunov exponent function. This type of relation was first established by Thouless, in [67], for Schrödinger cocycles in the Anderson model (see Example 2.9). For more general affine families locally constant cocycles, we have the following result.

Theorem 2.10 (Dynamical Thouless Formula [13] and [29]) *Let $A \in SL_2(\mathbb{R})^\kappa$. Assume that there exists a vector of matrices $(B_1, \dots, B_\kappa) \in SL_2(\mathbb{R})^\kappa$ such that*

1. *$rank(B_j) = 1$ and $A_j(Ker(B_j)) = Im(B_j)$;*
2. *For every $i, j = 1, \dots, \kappa$, $Im(B_i) \neq Ker(B_j)$.*

Then, the affine family $\{A_t := A + tB\}_{t\in\mathbb{R}} \subset SL_2(\mathbb{R})^\kappa$ satisfies, for all $t \in \mathbb{C}$,

$$L(A_t) = L(B) + \int_{-\infty}^{\infty} \log|t - s|\, d\rho(s). \tag{13}$$

Remark 2.8 The assumption 1 guarantees that indeed, $A_t \in SL_2(\mathbb{R})^\kappa$, for every $t \in \mathbb{R}$. Assumption 2 ensures that the family $\{A_t\}_{t\in\mathbb{R}}$ is positively winding and that $L(B) > -\infty$.

The statement of Theorem 2.10 presented here, is a particular case of the result in [13]. The result [29] also provides a good description of the fibered rotation number ρ as the Laplacian of the Lyapunov exponent in the sense of distributions.

The basic idea explored in the proof of Theorem 2.10 is the following: it is possible to recover the Lyapunov exponent of the cocycle A_t using the expression

$$L(A_t) = \max_{i,j=1,2} \{ \lim_{n\to\infty} \frac{1}{n} \log|\langle A_t^n(x)\, u_i,\, u_j\rangle|\},$$

where $\{u_1, u_2\}$ is any basis of $\mathbb{R}^2$. Notice that for each $n \in \mathbb{N}$ and each $x \in \Sigma$ the function $t \mapsto \langle A_t^n(x)\, u_i,\, u_j\rangle$ is a polynomial of degree n (with real roots) and leading coefficient given by $\langle B^n(x)\, u_i,\, u_j\rangle$. So, factoring this polynomial through its roots we have

$$\lim_{n\to\infty} \frac{1}{n} \log|\langle A_t^n(x)\, u_i,\, u_j\rangle| = \lim_{n\to\infty} \frac{1}{n} |\langle B^n(x)\, u_i,\, u_j\rangle| + \lim_{n\to\infty} \frac{1}{n} \sum_{t^*} \log|t - t^*|.$$

Choosing the basis $\{u_1, u_2\}$ appropriately such that the first term on the right-hand side above converges a.s. to $L(B)$, gives us that for any $i, j = 1, 2$,

$$\lim_{n\to\infty} \frac{1}{n} \log|\langle A_t^n(x)\, u_i,\, u_j\rangle| = L(B) + \lim_{n\to\infty} \frac{1}{n} \sum_{t^*} \log|t - t^*|.$$

Now, using the winding property we see that the second term in the right-hand side above converges to the desired integral.

Notice that the formula in (13) gives a relation between the regularity of the function $t \mapsto L(A_t)$ and the regularity of the fibered rotation number ρ. Indeed, the integral on the right-hand side of the formula can be rewritten, using integration by parts, as the Hilbert transform of ρ. By a result of Goldstein and Schlag [40, Lemma 10.3] the Hilbert transform preserves the Hölder modulus of continuity. Therefore, the following result is a consequence of Theorems 2.9, 2.10 and item 4 of Proposition 2.12.

Theorem 2.11 (Bezerra et al. [13]) *Assume that $A \in SL_2(\mathbb{R})^\kappa$ is not uniformly hyperbolic and $L(A) > 0$. If $\theta > H(\mu)/L(A)$, then L is not θ-Hölder continuous in a neighborhood of A is $SL_2(\mathbb{R})^\kappa$.*

We now make a few comments about the proof of Theorem 2.9. Naively speaking, the presence of a tangency allows us to produce sequences $x \in \Sigma$, called *matchings* of size k, for any k sufficient large, such that moving the parameter t inside a small interval I_k of size $e^{-k\,(L(A)-\varepsilon)}$ centered at t_0 the projective curve $I_k \ni t \mapsto A^k(x)\,\hat{v}$ goes one full circle around $\mathbb{P}^1$. In particular, $\ell_{I_k}(A^k(x)\hat{v}) \geq \pi$. This behavior is additive in the sense that factoring $A^{km}(x) = \prod_j A^k(\sigma^{jk}(x))$, the numbers of full circles that the block curves $I_k \ni t \mapsto A^k(\sigma^{jk}(x))\hat{v}$ go around $\mathbb{P}^1$ essentially add up. Hence

$$\frac{1}{km\pi}\ell_{I_k}(A^{km}(x)) \geq \frac{1}{km\pi}\sum_{j=0}^{k-1}\ell_{I_k}(A^k(\sigma^{jk}(x))\,\hat{v}) \geq \frac{1}{km\pi}\sum_{j=0}^{k-1}\mathcal{X}_{\mathcal{M}(n,J)}(\sigma^{jk}(x)),$$

where $\mathcal{M}(k, J) \subset \Sigma$ is the set of matchings of size k in the interval J (definition below). Therefore, we may use Birkhoff's ergodic theorem to guarantee that

$$\frac{\rho(I_k)}{|I_k|^\theta} \geq \frac{1}{k\pi}\frac{\mu(\mathcal{M}(k, J))}{|I_k|^\theta}. \tag{14}$$

In particular, if the right-hand side (RHS) of the above expression explodes, we obtain that ρ cannot be θ-Hölder continuous. So, a precise study of the set of matchings is required in order to give a lower bound for its probability.

Formally, we say that a sequence $x \in \Sigma$ is a *γ-matching* of size k at t_0 if there exist directions $\hat{v}, \hat{w} \in \mathbb{P}^1$ and a natural number $1 \leq m \leq k$ such that

1. $A^k_{t_0}(x)\,\hat{v} = \hat{w}$;
2. $e^\gamma \leq \left\|A^m_{t_0}(x)\,\hat{v}\right\| \leq \left\|A^m_{t_0}(x)\right\| \leq 2e^\gamma$;
3. $e^\gamma \leq \left\|A^{-(k-m)}_{t_0}(\sigma^k(x))\,\hat{w}\right\| \leq \left\|A^{-(k-m)}_{t_0}(\sigma^k(x))\right\| \leq 2e^\gamma$;

(See Fig. 3). In this case, we say that the pair of matrices (B, D), where $B = A^m(x)$ and $D = A^{k-m}(\sigma^m(x))$, connects $\hat{v}$ and $\hat{w}$ and produces the γ-matching (the value used for γ is $\gamma = k\,(L(A) - \varepsilon)$ for some conveniently small ε).

If (B, T, D) is a *balanced tangency* for A_{t_0}, i.e., both matrices B and D correspond to hyperbolic words with eigenvalues $\sim e^{\pm c}$, then we can easily

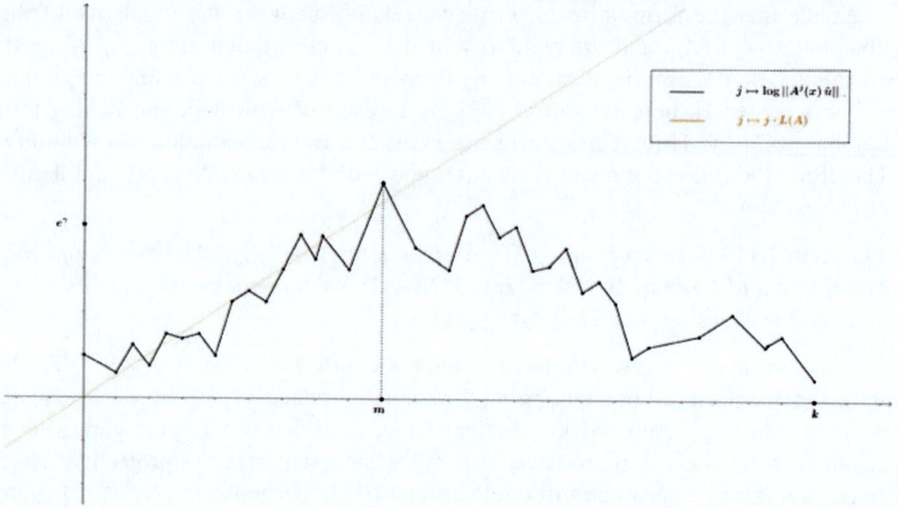

Fig. 3 Growth of the map $j \mapsto \log \left\| A^j(x)\, \hat{u} \right\|$ for a γ-matching

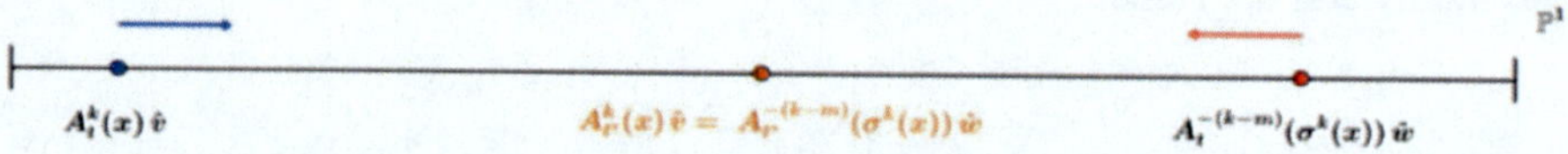

Fig. 4 Moving the parameter t to obtain matchings

find many other matchings for A_{t_0}. Indeed, the pair $(T\, B^m,\ D^m)$ produces a mc-matching for every $m \geq 1$. However, this way of finding matchings only gives us a set of zero probability, which is not suitable for our purposes.

To obtain matchings notice that for positively winding families, picking any directions $\hat{v}, \hat{w} \in \mathbb{P}^1$, the points $A_t^m(x)\, \hat{v}$ and $A_t^{-(k-m)}(\sigma^k(x))\, \hat{w}$ move (w.r.t. t) in *opposite directions* at a speed that is uniformly bounded away from zero. Assuming that the norms of $A_t^k(x)$ and $A_t^{-(k-m)}(\sigma^k(x))\, \hat{w}$ are large (which can be provided by the uniform large deviation estimates) this process eventually gives us a matching at some moment $t^* \in \mathbb{R}$, i.e., (see Fig. 4).

$$A_{t^*}^k(x)\, \hat{v} = A_{t^*}^{-(k-m)}(\sigma^k(x))\, \hat{w}.$$

To ensure that t^* occurs in a small neighborhood of t_0 the strategy is to use a tangency to force that at the initial time t_0 the points $A_{t_0}^k(x)\, \hat{v}$ and $A_{t_0}^{-k}(\sigma^k(x))\, \hat{w}$ are already close to each other.

Let τ be a finite word associated to a *balanced heteroclinic tangency* (B, T, D) with $A_{t_0}^{|\tau|}(y) = D\, T\, B$ for some $y \in [0; \tau]$. More precisely let $|\tau| = m_B + m_D + m_T$ where $B = A_{t_0}^{m_B}(y)$, $T = A_{t_0}^{m_T}(\sigma^{m_B}(y))$ and $D = A_{t_0}^{m_D}(\sigma^{m_B+m_T}(y))$. The size of the neighborhood I where we will look for matchings is determined by the *Lyapunov exponent of the tangency* τ

$$L(\tau) := \frac{1}{|\tau|} \log \left\| A_{t_0}^{|\tau|}(y) \right\| \approx \frac{1}{m_D} \log \|D\| \approx \frac{1}{m_B} \log \|B\|$$

in the sense that $|I| \sim e^{-|\tau|(L(\tau)-\varepsilon))}$ for some conveniently and arbitrarily small $\varepsilon > 0$. On the other hand, the measure of the matching occurrence event $\mathcal{M}(|\tau|, I)$ will be determined by the *entropy of the tangency* τ:

$$H(\tau) := -\frac{1}{|\tau|} \log \mu([0; \tau]),$$

in the sense that $\mu(\mathcal{M}(|\tau|, I)) \geq e^{-|\tau|(H(\tau)-\varepsilon)}$ for some small $\varepsilon > 0$.

The process to create multiple matchings goes as follows. Fix τ as before and take any points $\hat{u}, \hat{w} \in \mathbb{P}^1$. Consider matrix products B_n and D_n associated with much longer words of size $n \gg |\tau|$. For most choices of these words we will have

1. $\|B_n\, u\| \gtrsim e^{n\,(L(A)-\varepsilon)}$, $\left\|D_n^{-1}\, w\right\| \gtrsim e^{n\,(L(A)-\varepsilon)}$,
2. $B_n\, \hat{u}$ is at some distance bounded away from the stable eigen-direction of B,
3. $D_n^{-1}\, \hat{w}$ is at a distance bounded away from the unstable eigen-direction of D.

Then the pairs $(T\, B\, B_n,\ D_n\, D)$ are almost matchings at t_0 because

$$d(T\, B\, B_n\, \hat{u}, D^{-1}\, D_n^{-1} \hat{w}) \lesssim e^{-|\tau|(L(\tau)-\varepsilon)}.$$

See Fig. 5 for a graphical explanation of this process of creation of matchings.

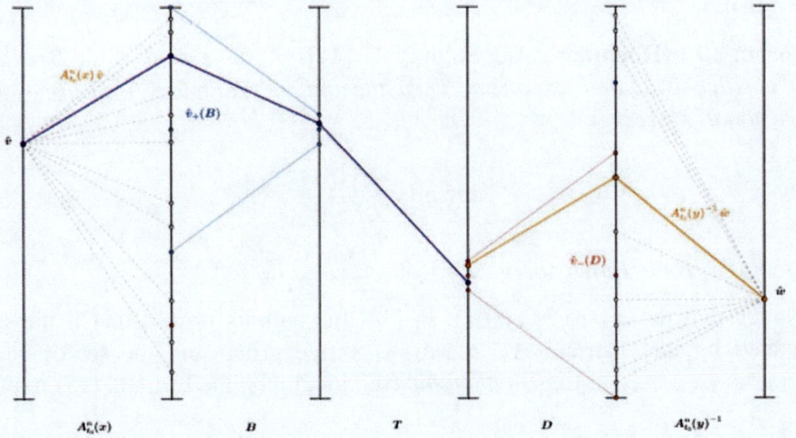

Fig. 5 Construction of matchings. The vertical bars represent different copies of the projective space and between them the respective matrix action is depicted

True matchings occur at nearby parameters t_n^* with $|t_n^* - t_0| \lesssim e^{-|\tau|(L(\tau)-\varepsilon)}$, in the sense that the pair $(T\ B\ B_n,\ D_n\ D)$ will have a matching occurring at t_n^*. In particular, the right-hand side of (14) is bounded from below by $e^{-|\tau|(H(\tau)-\theta\,L(\tau)-2\varepsilon)}$.

Since we are able to produce tangencies τ such that $|\tau| \to \infty$, if we could prove that $\theta > H(\tau)/L(\tau)$ for every such tangency, then $\theta > (H(\tau)-\varepsilon)/(L(\tau)-\varepsilon)$ for ε small enough, and we would conclude our result.

To overcome the possibility that the Lyapunov exponent and entropy of the tangency could be very different from $H(\mu)$ and $L(A_{t_0})$ we can produce many *typical tangencies* of arbitrarily large size, i.e., tangencies τ where $(H(\tau), L(\tau))$ is arbitrarily close to $(H(\mu), L(A_{t_0}))$, by essentially the same process that we are using to produce matchings. This concludes the sketch of the proof of Theorem 2.9.

2.5 *Regularity and Dimension of the Stationary Measures*

The relation between the entropy and the Lyapunov exponent obtained in the bound for the regularity of L in Theorem 2.11 resembles Ledrappier-Young type of formulas (see [52] and [53]). In the case of linear cocycles this quocient is related with the dimension of the forward and backward stationary measures.

The definition of upper and lower local dimensions of a projective probability measure η goes as follows:

$$\overline{\dim}\,(\eta;\ \hat{v}) = \overline{\lim_{r\to 0^+}}\ \frac{\log \eta(B_r(\hat{v}))}{\log r} \quad \text{and} \quad \underline{\dim}\,(\eta;\ \hat{v}) = \underline{\lim_{r\to 0^+}}\ \frac{\log \eta(B_r(\hat{v}))}{\log r}.$$

We say that η is *exact dimensional* if there exists a number $\alpha \geq 0$ such that for η-a.e. every $\hat{v} \in \mathbb{P}^1$ we have $\overline{\dim}\,(\eta;\ \hat{v}) = \underline{\dim}\,(\eta;\ \hat{v}) = \alpha$. In this case we write $\alpha = \dim \eta$.

Theorem 2.12 (Hochman and Solomyak [44]) *Assume that A is irreducible with positive Lyapunov exponent. Then the unique stationary measure is exact dimensional. Moreover,*

$$\dim \eta = \frac{h_A(\eta)}{2\,L(A)}, \tag{15}$$

where $h_A(\eta)$ is the Furstenberg entropy introduced in (6).

Notice that the $h_A(\eta) \leq H(\mu)$. In [44] the authors proved that if the group generated by the matrices $A_1, \ldots, A_\kappa$ is a free group and the set of cocycle generators $\{A_1, \ldots, A_\kappa\}$ is *Diophantine* (see [44]) then the formula (15) improves to

$$\dim \eta^+ = \dim \eta^- = \frac{H(\mu)}{2\,L(A)},$$

where η^+ and η^- denotes respectively the forward and backward stationary measure of the cocycle A. So, the upper bound for the regularity of the Lyapunov exponent is in these cases is nothing more than $\dim \eta^+ + \dim \eta^-$.

Problem 2.1 *Is* $\dim \eta^+ + \dim \eta^-$ *always an upper bound for the regularity of the Lyapunov exponent?*

A related problem concerns the symmetry of the forward and backward systems.

Problem 2.2 *Is there an example of irreducible cocycle with* $L(A) > 0$ *where* $\dim \eta^+ \neq \dim \eta^-$*?*

We could also wonder what is the lower bound for the regularity of typical cocycles. For a cocycle $A \in SL_2(\mathbb{R})^\kappa$ define the quantity

$$\theta_A = \sup\{\theta \geq 0\colon A \text{ is } \theta - \text{Hölder around } A\}.$$

Problem 2.3 *Assume* A *irreducible with* $L(A) > 0$. *What is the natural lower bound for* θ_A*?*

As we have seen in the discussion of the proof of Theorem 2.6, the quantity θ_A is associated to the contraction properties of the Markov operator. This could be the path to obtain an interesting answer to problem 2.3.

Remark 2.9 The topics discussed above for locally constant cocycles, i.e., maps $A : \Sigma \to SL_2(\mathbb{R})$ that depend only on the zero-th coordinate, can be generalized to maps that depends on a finite (but fixed, say m) number of coordinates. Indeed such cocycles can always be regarded as locally constant maps over a full shift in κ^m symbols.

A full understanding of the Lyapunov exponent function among the locally constant is far from complete. One example of the lack of precise knowledge about this function is given by the next open problem

Problem 2.4 *Is there an example of locally constant cocycle which is a* C^r*-continuity point of the Lyapunov exponent for some* $r \geq 1$*?*

Remark 2.10 We would expect that such a cocycle admits a very regular stationary measure. One example of a cocycle with C^r, for r sufficiently large, stationary measure was provided by J. Bourgain [23]. See [12] for a better discussion on the regularity of stationary measures in the context of group actions.

3 Holder Random Cocycles

Now we go back to the initial discussion for continuous linear cocycles $A : \Sigma \to SL_2(\mathbb{R})$ possibly depending on infinitely many coordinates. To fully extract the hyperbolic properties of the basic dynamics it will be convenient to establish some notation.

For each $x = (x_n)_{n\in\mathbb{Z}} \in \Sigma$ we may write $x = (x^-, x^+)$, where $x^- = (x_{-n})_{n\leq 0}$ and $x^+ = (x_n)_{n\geq 0}$. We define the local stable and local unstable set of x respectively by

$$W^s_{\text{loc}}(x) = \{y \in \Sigma\colon x^+ = y^+\} \quad \text{and} \quad W^u_{\text{loc}}(x) = \{y \in \Sigma\colon x^- = y^-\}.$$

For each $x, y \in \Sigma$ with $x_0 = y_0$, $W^s_{\text{loc}}(x)$ and $W^u_{\text{loc}}(y)$ intersect each other at a single point, namely, $[x, y] := (y^-, x^+)$.

We consider the space $C^\gamma(\Sigma, SL_2(\mathbb{R}))$ endowed with the γ-Hölder topology C^γ: $A \in C^\gamma(\Sigma, SL_2(\mathbb{R}))$, $\|A\|_{C^\gamma} = \|A\|_\infty + [A]_\gamma$, where

$$[A]_\gamma = \sup_{x,y\in\Sigma} \frac{\|A(x) - A(y)\|}{d(x, y)^\gamma},$$

and d denotes is the usual distance on Σ,

$$d(x, y) := \zeta^{-N(x,y)}, \qquad \text{with } N(x, y) = \inf\{|n|\colon x_n \neq y_n\}, \qquad \zeta > 1. \tag{16}$$

Take $A \in C^\gamma(\Sigma, SL_2(\mathbb{R}))$ and let $F_A : \Sigma\times\mathbb{P}^1 \to \Sigma\times\mathbb{P}^1$ be the projective linear cocycle associated with A. Let $\mathcal{M}_\mu(F_A)$ be the space of F_A invariant probability measures that project on μ in the first coordinate.

The analysis of F_A-invariant measures in the study of Lyapunov exponent for cocycles, now depending on infinitely many coordinates, plays a very similar role to the analysis of stationary measures for locally constant cocycles. So, we start listing a few properties of this class of measures.

Proposition 3.1 *It holds that,*

1. *$\mathcal{M}_\mu(F_A)$ is non-empty, compact and convex. The extremal points of $\mathcal{M}_\mu(F_A)$ consist of ergodic measures for F_A;*
2. *For each $m \in \mathcal{M}_\mu(F_A)$, there exists an (essentially unique) measurable family of measures $\{m_x\}_{x\in\Sigma}$ on $\mathbb{P}^1$, the disintegration of m, such that for every measurable product $X \times \hat{V} \subset \Sigma \times \mathbb{P}^1$,*

$$m(X \times \hat{V}) = \int_X m_x(\hat{V})\, d\mu(x).$$

Moreover, $A(x)_* m_x = m_{\sigma(x)}$, for μ almost every $x \in \Sigma$;

3. *Assume that for every $x \in \Sigma$, $A(x) = A(x^+)$, i.e., A only depends on the positive coordinates. The projection of an (ergodic) measure $m \in \mathcal{M}_\mu(F_A)$ on $\Sigma^+\times\mathbb{P}^1$ is an (ergodic) probability measure invariant by F_A^+ (the natural map). Conversely, if m^+ is an (ergodic) F_A^+-invariant probability measure on $\Sigma^+ \times \mathbb{P}^1$ projecting on μ, then there exists a unique (ergodic) measure $m \in \mathcal{M}_\mu(F_A)$ such that the*

projection is m^+. The disintegration of m and m^+ are related by

$$m^+_{x^+} = \int_{W^s_{loc}(x^+)} m_x \, d\mu^s_{x^+}(x), \quad m_x = \lim_{n\to\infty} A^n(\sigma^{-n}(x))\, m^+_{\sigma^{-n}(x)^+},$$

where $\{\mu^s_{x^+}\}_{x^+\in\Sigma^+}$ is the disintegration of μ with respect to the partition on local stable sets. A similar property holds when A only depends on the negative coordinates;

4. *If $L(A) > 0$, denote by $\hat{e}^s$, $\hat{e}^u$ the Oseledets directions (Proposition 1.1). Write*

$$m^u = \int \delta_{\hat{e}^u}\, d\mu \quad \text{and} \quad m^s = \int \delta_{\hat{e}^s}\, d\mu.$$

Then, $[m^u, m^s] = \mathcal{M}_\mu(F_A)$, i.e., any $m \in \mathcal{M}_\mu(F_A)$ can be written as

$$m = q_1 m^u + q_2 m^s = \int q_1 \delta_{\hat{e}^u} + q_2\, \delta_{\hat{e}^s}\, d\mu.$$

with $q_1 \geq 0$, $q_2 \geq 0$, $q_1 + q_2 = 1$.

Outline of the Proof and References Item 1 follows a similar strategy as in the proof of item 1 in Proposition 2.1. Item 2 is an application of Rokhlin's disintegration theorem with respect to the partition $\{\{x\} \times \mathbb{P}^1 : x \in \Sigma\}$ of $\Sigma \times \mathbb{P}^1$. The invariance of the conditional measures m_x comes from the invariance of m by F_A. For a proof of item 3 see [7, Proposition 3.4]. Item 4 can be found in [69, Lemma 5.25]. □

3.1 Fiber Bunched Cocycles

As mentioned before, the complete description of the regularity of the Lyapunov exponent among cocycles in $C^\gamma(\Sigma, SL_2(\mathbb{R}))$ is still open, but there is a class of cocycles where significant progress has been made. This is the class of maps A such that the growth rate of the quantities $\|A^n\|$ occurs slower than the expansion provided by the base dynamics. More precisely, we say that a cocycle $A \in C^\gamma(\Sigma, SL_2(\mathbb{R}))$ is *γ-fiber bunched* if there exists $n \in \mathbb{N}$ such that

$$\sup_{x\in\Sigma} \left\|A^n(x)\right\|^2 = \sup_{x\in\Sigma} \left\|A^n(x)\right\| \left\|(A^n(x))^{-1}\right\|^{-1} < \zeta^{n\gamma},$$

where $\zeta > 1$ appears in the definition (16) of the distance on Σ. It is important to highlight that the class of γ-fiber bunched cocycles $FB^\gamma(\Sigma)$ is an open subset of $C^\gamma(\Sigma, SL_2(\mathbb{R}))$.

Remark 3.1 A direct computation shows that the derivative of the projective actions $A^n(x) : \mathbb{P}^1 \to \mathbb{P}^1$ satisfies that, for every $\hat{w} \in \mathbb{P}^1$

$$\left\| D\left[A^n(x)\right](\hat{w}) \right\| \leq \frac{1}{\|A^n(x)\, w\|^2} \leq \left\| A^n(x) \right\|^2. \tag{17}$$

Thus, assuming A γ-fiber bunched, we have

$$\sup_{x\in\Sigma} \left\| D[A^n(x)] \right\|_\infty \leq \zeta^{n\gamma}, \tag{18}$$

In particular, the system $F_A : \Sigma \times \mathbb{P}^1 \to \Sigma \times \mathbb{P}^1$ can be seen as a basic model of a *partial hyperbolic system* (with compact fibers). Indeed, if we interpret $\sigma : \Sigma \to \Sigma$ as the basic model for hyperbolic system, then Eq. (18) means that the fiber bunched condition guarantees that the growth rate of the fiber action $A^n(x)$ is controlled by the growth rate of the hyperbolic basis σ.

Example 3.1 (Bocker-Viana's Example) Take $\Sigma = \{1, 2\}^{\mathbb{Z}}$, i.e., $\kappa = 2$. For any $\beta \geq \alpha \geq 1$ consider the map $A_{\alpha,\beta} : \Sigma \to SL_2(\mathbb{R})$ that for each sequence $x = (x_n)_{n\in\mathbb{Z}}$ is given by

$$A_{\alpha,\beta}(x) = \begin{cases} A_1 = \begin{pmatrix} \beta & 0 \\ 0 & \beta^{-1} \end{pmatrix}, & x_0 = 1 \\ A_2 = \begin{pmatrix} \alpha^{-1} & 0 \\ 0 & \alpha \end{pmatrix}, & x_0 = 2 \end{cases}$$

Thus, it is clear that $A_{\alpha,\beta} \in C^\gamma(\Sigma,\ SL_2(\mathbb{R}))$ ($A_{\alpha,\beta}$ is constant into cylinders of the form $[0; i]$, $i = 1, 2$). Notice that $\sup_{x\in\Sigma} \left\| A^n_{\alpha,\beta}(x) \right\|^2 = \beta^{2n}$ and so, $A_{\alpha,\beta}$ is γ-fiber bunched as long as $\beta < \zeta^{\gamma/2}$. Observe that this class of cocycles is not uniformly hyperbolic. Moreover, by Birkhoff's ergodic theorem,

$$L(A_{\alpha,\beta}) = |p_1 \log\beta - p_2 \log\alpha|.$$

and so $L(A_{\alpha,\beta}) > 0$, except for the case where $\beta^{p_1} = \alpha^{p_2}$.

3.2 *Holonomies*

Before discussing the main results of this section, we elaborate on one of the most important features of fiber bunched cocycles, namely, the existence of linear holonomies. We say that a family of matrices $\{H^s_{x,y} \in SL_2(\mathbb{R}){:}\ x, y \in \Sigma,\ y \in W^s_{\text{loc}}(x)\}$ is a family of *stable linear holonomies* if

1. $H^s_{\sigma^j(x),\sigma^j(y)} = A^j(y) \circ H^s_{x,y} \circ A^j(x)^{-1}$ for every $j \geq 1$;

2. $H^s_{x,x} = \mathrm{Id}$ and $H^s_{x,y} = H^s_{z,y} \circ H^s_{x,z}$ for any $z \in W^s_{loc}(x)$;
3. $(x, y) \mapsto H^s_{x,y}$ is uniformly continuous on $\{(x, y) : y \in W^s_{\mathrm{loc}}(x)\} \subset M \times M$.

The family of *unstable linear holonomies* $\{H^u_{x,y} \in SL_2(\mathbb{R}) \colon y \in W^u_{\mathrm{loc}}(x)\}$ is defined analogously.

Remark 3.2 Existence of families of stable and unstable holonomies for A is also related with the existence of strong stable/unstable sets for the (partially hyperbolic in the fiber bunched case) skew-product map $F_A : \Sigma \times \mathbb{P}^1 \to \Sigma \times \mathbb{P}^1$: for every $x \in \Sigma$, $\hat{v} \in \mathbb{P}^1$, define

$$W^{ss}(x, \hat{v}) = \left\{(y,\ H^s_{x,y}\,\hat{v})\colon\ y \in W^s_{\mathrm{loc}}(x)\right\} \quad \text{and}$$
$$W^{uu}(x, \hat{v}) = \left\{(y,\ H^u_{x,y}\,\hat{v})\colon\ y \in W^u_{\mathrm{loc}}(x)\right\}.$$

Similarly to what happens for smooth partially hyperbolic systems, the map F_A restricted to these stable/unstable sets satisfy contraction/expansion asymptotic properties (see [68, Lemma 2.10] for a precise statement).

Example 3.2 Assume that $A \in C^\gamma(\Sigma, SL_2(\mathbb{R}))$ is a diagonal cocycle, i.e.,

$$A(x) = \begin{pmatrix} a(x) & 0 \\ 0 & a(x)^{-1} \end{pmatrix}.$$

Consider $x, y \in \Sigma$, $y \in W^s_{\mathrm{loc}}(x)$. For every $n \geq 1$,

$$A^n(y)^{-1}\, A^n(x) = \begin{pmatrix} \varphi_n(x, y) & 0 \\ 0 & \varphi_n(x, y)^{-1} \end{pmatrix} \quad \text{with } \varphi_n(x, y) = \frac{a(x)}{a(y)} \cdots \frac{a(\sigma^{n-1}(x))}{a(\sigma^{n-1}(y))}.$$

We claim that the limit $\varphi(x, y) := \lim_{n\to\infty} \varphi_n(x, y)$ exists and so using this claim,

$$H^s_{x,y} := \begin{pmatrix} \varphi(x, y) & 0 \\ 0 & \varphi(x, y)^{-1} \end{pmatrix}.$$

defines a family of stable holonomies for A (the conditions in the definition can be easily verified). To prove the claim, first consider $C > 0$ such that for every $z, w \in \Sigma$,

$$|\log a(z) - \log a(w)| \leq C d(z, w)^\gamma.$$

Then, since $y \in W^s_{\mathrm{loc}}(x)$

$$\begin{aligned} |\log \varphi_{n+1}(x, y) - \log \varphi_n(x, y)| &= |\log a(\sigma^n(x)) - \log a(\sigma^n(y))| \\ &\leq C\, d(\sigma^n(x), \sigma^n(y))^\gamma \leq C\, \zeta^{-n\gamma}. \end{aligned}$$

So, $(\varphi_n(x, y))_n$, is a (uniform) Cauchy sequence. This proves the claim.

Example 3.3 Let $\xi : \Sigma \to \mathbb{R}$ be a γ-Hölder function and define $A(x) = R_{2\pi\xi(x)}$. For $x, y \in \Sigma$, $y \in W^s_{\mathrm{loc}}(x)$ and for every $n \geq 1$,

$$A^n(y)^{-1} A^n(x) = R_{S_n\xi(x)-S_n\xi(y)},$$

where $S_n\xi$ denotes the Birkhoff's sum of the function ξ. We claim that $(S_n\xi(x) - S_n\xi(y))_n$ forms a Cauchy sequence. Indeed, to see that is enough to notice that

$$|S_{n+1}\xi(x)-S_{n+1}\xi(y) - (S_n\xi(x)-S_n\xi(y))| \leq |\xi(\sigma^n(x))-\xi(\sigma^n(y))| \leq C\,\zeta^{-n\gamma}.$$

Now, let $\xi_0(x, y) = \lim_{n\to\infty} (S_n\xi(x) - S_n\xi(y))$. Then, the $H^s_{x,y} := R_{\xi_0(x,y)}$ defines a family of s-holonomies for A (the properties in the definition are easily verified).

Remark 3.3 Notice that the cocycle A in the Example 3.3 is always γ-fiber bunched since it is bounded. On the other hand, Example 3.2 may not be γ-fiber bunched, but can always be seen as a "direct product of fiber bunched" 1-dimensional cocycles. Indeed, in this case the existence of the limits $A^n(y)^{-1} A^n(x)$ is ensured by the conformality of the maps $A|_{\hat{e}_1}$ and $A|_{\hat{e}_2}$.

Proposition 3.2 *If $A \in C^\gamma(\Sigma, SL_2(\mathbb{R})^\kappa$ is γ-fiber-bunched, then the families*

- $\{\lim_{n\to\infty} A^n(y)^{-1} A^n(x)\colon\ y \in W^s_{loc}(x)\}$;
- $\{\lim_{n\to\infty} A^n(\sigma^{-n}(y))\, A^n(\sigma^{-n}(x))^{-1}\colon\ y \in W^u_{loc}(x)\}$

are respectively the unique family of stable and unstable linear holonomies such that for some $L > 0$

$$\|H^s_{x,y} - \mathrm{Id}\,\| \leq L\,\mathrm{dist}(x, y)^\gamma \quad \textit{and} \quad \|H^u_{x,y} - \mathrm{Id}\,\| \leq L\,\mathrm{dist}(x, y)^\gamma. \tag{19}$$

Outline of the Proof and References For a proof see [19, Lemma 1.12]. □

Remark 3.4 In general, the family of stable and unstable holonomies are not unique even for a constant fiber bunched cocycle (see the discussion after 4.9 in [47] and Theorem 5.5.5 in [48]). However, families of holonomies satisfying the conditions in (19) are unique even without fiber bunched assumptions (See Proposition 3.2 in [47]).

Remark 3.5 Any $C^\gamma(\Sigma, SL_2(\mathbb{R}))$ cocycle , C^0-close to a constant one, has families of stable and unstable linear holonomies satisfying (19), possibly with some $\gamma' < \gamma$ instead of γ (see [47, Proposition 3.6]).

Example 3.4 If $A \in SL_2(\mathbb{R})^\kappa$ is locally constant then the identity map forms a family of stable/unstable holonomies. Assume that there exists another family of linear stable holonomies, say $\{H^s_{x,y}\}$. Then, for every $x, y \in \Sigma$, $y \in W^s_{\mathrm{loc}}(x)$ notice that $A^n(x) = A^n(y)$ and so we have

$$H^s_{\sigma^n(x),\sigma^n(y)} = A^n(x)\, H^s_{x,y}\, A^n(x)^{-1}.$$

In particular, $H^s_{\sigma^n(x),\sigma^n(y)}$ and $H^s_{x,y}$ share the same spectrum for every n ($A^n(x)$ is a conjugation). Since $d(\sigma^n(x), \sigma^n(y))$ converges to zero, we have that the spectrum of the matrix $H^s_{x,y}$ is a singleton with only 1. So, either $H^s_{x,y}$ is the identity matrix or else is conjugated to a parabolic matrix of the form $\begin{pmatrix} 1 & a \\ 0 & 1 \end{pmatrix}$.

One of the most important uses of the existence of holonomies is the possibility to study the problem from the point of view of one-sided shift.

Proposition 3.3 *If A admits a family of linear stable holonomies, then A is conjugated to a cocycle A^+ such that for every $x \in \Sigma$, $A^+(x) = A^+(x^+)$. Similar property holds for unstable linear holonomies.*

Proof See [19, Corollary 1.15]. □

Remark 3.6 One could think that once the cocycle admits both families of holonomies, then it would be conjugated to a locally constant cocycle. Unfortunately, this is not the case in general. Being conjugate to a locally constant cocycle is related to the problem of accessibility of partially hyperbolic systems.

A map $F_A : \Sigma \times \mathbb{P}^1 \to \Sigma \times \mathbb{P}^1$ is said to be accessible if for every $(x, \hat{v}), (y, \hat{u}) \in \Sigma \times \mathbb{P}^1$ there exists a path of strong stable and strong unstable sets, introduced in Remark 3.2, that goes from $(x, \hat{v})$ to $(y, \hat{u})$.

Stable and unstable sets of a cocycle are mapped to the corresponding ones for the conjugated cocycle via the conjugation map. This implies that accessibility is preserved by conjugation. However, locally constant cocycles are never accessible. This shows that γ-fiber bunched accessible maps can not be conjugated to locally constant cocycles even though can be conjugated to cocycles depending only on the positive (or negative) coordinates.

3.3 Su-States

Holonomies allows us to transport components of the disintegration $\{m_x\}_{x\in\Sigma}$ of a given measure $m \in \mathcal{M}_\mu(F_A)$ throughout different points in the same stable set. Taking $x, y \in \Sigma$, $y \in W^s_{\text{loc}}(x)$, we are able to compare the structure of m_y with the structure of $(H^s_{x,y})_* m_x$.

One especial case of this transport property stands out. We say that m is an *s-state* if for μ almost every $x, y \in \Sigma$, $y \in W^s_{\text{loc}}(x)$, $(H^s_{x,y})_* m_x = m_y$. Analogously, m is an *u-state* if for μ almost every $x, y \in \Sigma$, $y \in W^u_{\text{loc}}(x)$, $(H^u_{x,y})_* m_x = m_y$. m is an *su-state* when it is both an s-state and an u-state.

Example 3.5 Assume that $A \in SL_2(\mathbb{R})^\kappa$ is locally constant. Then, it is easy to see that the identity matrix forms a family of stable linear holonomies for A. In this case, a measure $m \in \mathcal{M}_\mu(F_A)$ is an s-state if an only if it has a disintegration $\{m_x\}_x$ which is constant along stable sets. In particular, the projection of m on $\Sigma^+ \times \mathbb{P}^1$ is the product measure $\mu \times \eta$, where η is a forward stationary measure for the cocycle A. Reciprocally, if η is a forward stationary measure, then the lift $\mu \times \eta$ (see Proposition 3.1) is constant along stable sets and so is an s-state.

The previous example indicates that s and u states are the natural generalization of the concept of stationary measure when dealing with cocycles which (possibly) depend on infinitely many coordinates.

Proposition 3.4 *Let $A \in C^\gamma(\Sigma, SL_2(\mathbb{R}))$ be a γ-fiber bunched cocycle.*

1. *The set of measures $m \in \mathcal{M}_\mu(F_A)$ which are s-states (respect. u-states) is non-empty, compact and convex;*
2. *If $L(A) > 0$, then the measure m^s is an s-state and m^u is an u-state;*
3. *Assume that $L(A) > 0$. Either m^s (respect. m^u) is the unique s-state (u-state) or else all the measures in $\mathcal{M}_\mu(F_A)$ are s-states (u-states);*
4. *Assume that $L(A) > 0$, we have (the following consequence of Oseledets Theorem—Proposition 1.1)*

$$L(A) = \int_{\Sigma \times \mathbb{P}^1} \log \|A(x)\, v\| \; d\, m^u(x, \hat{v}) = - \int_{\Sigma \times \mathbb{P}^1} \log \|A(x)\, v\| \; d\, m^s(x, \hat{v});$$

Outline of the Proof and References For a proof of item 1 see [21, Proposition 4.4]. Item 2 and 4 are consequences of Oseledets theorem (see item 4 of Proposition 1.1). For a proof of item 3, see [69, Corollary 5.27]. □

The next proposition shows the strength of the concept of su-states. It allows us to bypass the issues associated with the lack of regularity of the map $x \mapsto m_x$ which is only measurable in general.

Proposition 3.5 *Let $A \in C^\gamma(\Sigma, SL_2(\mathbb{R}))$ be a β fiber bunched cocycle.*

1. *If $m \in \mathcal{M}_\mu(F_A)$ is an u-state and $A(x) = A(x^+)$, then the projection of m, m^+, on $\Sigma^+ \times \mathbb{P}^1$ admits a Hölder continuous disintegration $x^+ \mapsto m^+_{x^+}$.*
2. *If $m \in \mathcal{M}_\mu(F_A)$ is a su-state, then m admits a continuous disintegration.*
3. *Assume that $L(A) > 0$. If there exists more than one s-state in $\mathcal{M}_\mu(F_A)$, then A is continuously conjugated to a triangular cocycle, i.e., there exists $C : \Sigma \to SL_2(\mathbb{R})$ continuous such that*

$$C(\sigma^{-1}(x))\, A(x)\, C(x) = \begin{pmatrix} a(x) & b(x) \\ 0 & d(x) \end{pmatrix}.$$

The same conclusion holds if there exists more than one u-state in $\mathcal{M}_\mu(F_A)$.

4. *Assume that $L(A) > 0$. If there exists a non-ergodic measure $m \in \mathcal{M}_\mu(F_A)$ which is an su-state, then A is continuously conjugated to a diagonal cocycle, i.e., there exists $C : \Sigma \to SL_2(\mathbb{R})$ continuous such that*

$$C(\sigma^{-1}(x))\, A(x)\, C(x) = \begin{pmatrix} a(x) & 0 \\ 0 & d(x) \end{pmatrix}.$$

Outline of the Proof and References For a proof of item 1 see [21, Proposition 4.3]. Item 2 is proved in [7, Proposition 4.8]. Items 3 and 4 are consequences of the item 2. □

The above proposition is an indication of the rigidity of the space of cocycles admitting su-states. These measures appear naturally when the Lyapunov exponent of the cocycle in consideration vanishes. This is the content of the next result.

Theorem 3.6 (Invariance Principle [7]) *If* $A \in C^\gamma(\Sigma, SL_2(\mathbb{R}))$ *is* γ*-fiber bunched and* $L(A) = 0$*, then any measure* $m \in \mathcal{M}_\mu(F_A)$ *is a su-state.*

Recall that for locally constant cocycles $A \in SL_2(\mathbb{R})^\kappa$ the invariance principle states, under the assumption $L(A) = 0$, that any stationary measure is fixed through the action of all the matrices A_i. Here, this invariance property is by the fact that both holonomies preserves the disintegrations of any measure $m \in \mathcal{M}_\mu(F_A)$. The proof of Theorem 3.6 follows similar lines as the proof of Theorem 2.2.

3.4 *Typical Fiber Bunched Cocycles*

The invariance principle provides a criterion to check positivity of the Lyapunov exponent: it is enough to find a single measure $m \in \mathcal{M}_\mu(F_A)$ which is not an su-state. Certainly, such a criterion is not very useful in practice. Our aim is to present a criterion for positivity of the Lyapunov exponent in the same lines as Furstenberg's criterion in the context of locally constant cocycles.

Let $A \in C^\gamma(\Sigma, SL_2(\mathbb{R}))$ be a cocycle admitting families of linear stable and unstable holonomies. One necessary condition to guarantee positivity the Lyapunov exponent of the cocycle A is the existence of some periodic point $q \in \Sigma$, of period k, which is a *pinching* periodic point, i.e., the matrix $A^k(q)$ is hyperbolic. In this case, we have two eigen-directions denoted by $\hat{v}_1(q), \hat{v}_2(q) \in \mathbb{P}^1$ for the matrix $A^k(q)$.

Assume for a moment that A admits an su-state $m \in \mathcal{M}_\mu(F_A)$. In particular, by continuity of the disintegration it makes sense to consider the projective measure m_q. Observe that, this is an invariant measure for the matrix $A^k(q)$ and so, it must to have the following form,

$$m_q = a\,\delta_{\hat{v}_1(q)} + b\,\delta_{\hat{v}_2(q)},$$

with $a, b \geq 0$, $a + b = 1$. Then, using the invariance of m by the holonomies and by the cocycle it is not hard to see that for any $z \in \Sigma$, there exist directions $\hat{v}_1(z), \hat{v}_2(z) \in \mathbb{P}^1$ such that

$$m_z = a\,\delta_{\hat{v}_1(z)} + b\,\delta_{\hat{v}_2(z)}.$$

Furthermore, the set of continuous sections $\{\hat{v}_1, \hat{v}_2 \in C^0(\Sigma, \mathbb{P}^1)\}$ is F_A-invariant in the sense that for every $z \in \Sigma$, $A(z)\,\hat{v}_i(z) \in \{\hat{v}_1(\sigma(z)), \hat{v}_2(\sigma(z))\}$ for $i = 1, 2$. So, if we want to guarantee that su-states are not allowed in our system we should ask that this type invariant set does not exists.

Now we introduce a concept incompatible with the existence of F_A-invariant sets of continuous sections. Let $z \in W^u_{\mathrm{loc}}(q)$ be a point homoclinic related to q meaning that $z^- = q^-$ and for some $l \geq 1$, $\sigma^l(z) \in W^s_{\mathrm{loc}}(q)$. Assume that the map $H^s_{\sigma^l(z),q}\, A^l(z)\, H^u_{q,z}$ *twists* the invariant subspaces of $A^k(q)$, meaning that

$$H^s_{\sigma^l(z),q}\, A^l(z)\, H^u_{q,z}\left(\{\hat{v}_1(q), \hat{v}_2(q)\}\right) \cap \{\hat{v}_1(q), \hat{v}_2(q)\} = \varnothing. \tag{20}$$

In particular, Eq. (20) implies that invariant sets of continuous sections as above do not exist.

The above discussion motivates the following definition: we say that the cocycle A is *pinching* if there exists a periodic point q as above such that $A^k(q)$ is hyperbolic, and we say that A is *twisting* if there exists a homoclinic point z related to q such that the property in (20) is satisfied.

A weaker form of twisting is sufficient for most of the results discussed in these notes. We call $\rho = \{x_0, \cdots, x_{n+1}\}$ an *su-loop* if $x_{i+1} \in W^*_{\mathrm{loc}}(x_i)$, for $* \in \{s, u\}$ and $i = 0, \ldots, n$ with $x_0 = x_{n+1}$. We say that a pinching cocycle is *weak twisting* if there exists an su-loop with $x_0 = q$ such that the map $H_\rho = H^{*_n}_{x_n,x_{n+1}} \circ \cdots \circ H^{*_0}_{x_0,x_1}$, where $*_i = s$ or u according to the definition of the loop, satisfies

$$H_\rho\left(\{\hat{v}_1(q), \hat{v}_2(q)\}\right) \cap \{\hat{v}_1(q), \hat{v}_2(q)\} = \varnothing. \tag{21}$$

Of course twisting implies weak twisting. Observe that the pinching and weak twisting condition also implies that there are no su-states (Fig. 6).

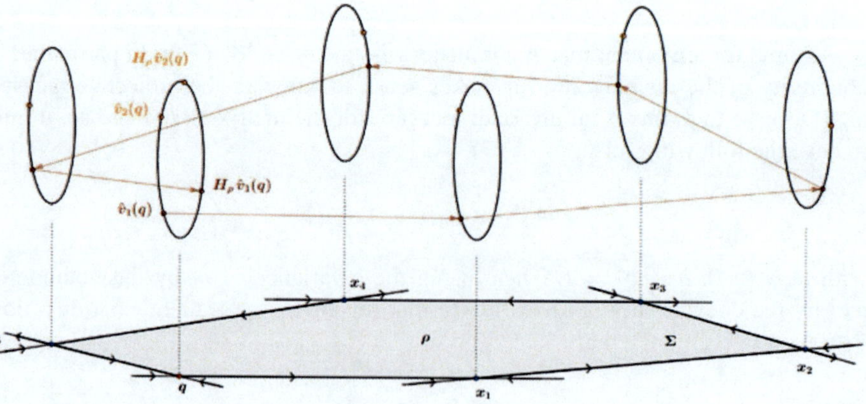

Fig. 6 Weak twisting

Problem 3.1 *Does weak twisting imply twisting?*

Remark 3.7 Note that when speaking about twisting property it is assumed that the cocycle in consideration admits families of stable and unstable holonomies.

The next proposition summarizes the above the discussion:

Proposition 3.7 *Let $A \in C^\gamma(\Sigma, SL_2(\mathbb{R}))$ be a γ-fiber bunched cocycle.*

1. *The set of pinching and twisting cocycles is an open and dense subset of $FB^\gamma(\Sigma)$;*
2. *If A is pinching and weak twisting, then there are no su-states;*
3. *If $L(A) > 0$, then A is pinching;*
4. *Assume that $L(A) > 0$. A is weak twisting if and only if there are no su-states;*
5. *If A is pinching and weak twisting, then $L(A) > 0$. Moreover, there is only one u-state and one s-state, namely m^u and m^s respectively;*
6. *Let A be a pinching and twisting cocycle depending only on the positive coordinates. If $m \in \mathcal{M}_\mu(F_A)$ is an u-state and m^+ is its projection on $\Sigma^+ \times \mathbb{P}^1$ with a continuous disintegration $\{m^+_{x^+}\}_{x^+}$, then for every $x^+ \in \Sigma^+$, $m^+_{x^+}(\{v\}) = 0$ for every $v \in \mathbb{P}^1$. A similar property holds for A depending only on the negative coordinates;*

Outline of the Proof and References For a proof of item 1 see [19, Theorem 7]. Item 2 is a consequence of the discussion just before the statement. For 3 see [46, Theorem 1.4].

One side of 4 is a direct consequence of items 2 and 3. Now, assume that $L(A) > 0$ and that there are no su-states. Let $q \in \Sigma$ be a σ-periodic given by item 3. For simplicity assume that $\sigma(q) = q$. Let $\hat{V} = \{\hat{v}_1, \hat{v}_2\} \subset \mathbb{P}^1$ be the set of invariant directions of the matrix $A(q)$.

If A is not twisting, then we have that,

$$H_\rho(\hat{V}) \cap \hat{V} \neq \varnothing \quad \text{for every } su\text{-loop } \rho. \tag{22}$$

If either there exists $\hat{v} \in \hat{V}$ such that $H_\rho\, \hat{v} = \hat{v}$ for every su-loop ρ, or $H_\rho(\hat{V}) = \hat{V}$, for every su-loop ρ we are able to build a su-state just considering the atomic measure with even weights in each case contradicting the assumption that there are no su-states. Therefore, we may assume that these cases do not happen. In particular, we can find a su-loop ρ_1 such that $H_{\rho_1}(\hat{v}_1) \notin \hat{V}$. By (22), we have that $H_{\rho_1}(\hat{v}_2) \in \hat{V}$. Consider the following cases:
Case 1: Assume that $H_{\rho_1}(\hat{v}_2) = \hat{v}_1$. In this case, there exists $n \in \mathbb{N}$ such that the map $H_{\rho_1} A^n(q) H_{\rho_1}$ (also associated to an su-loop) satisfies $H_{\rho_1} A^n(q) H_{\rho_1}(\hat{v}_1) \notin \hat{V}$, for $i = 1, 2$.
Case 2: Assume that $H_{\rho_1}(\hat{v}_2) = \hat{v}_2$. Observe that in this case there exists an su-loop ρ_2 such that $H_{\rho_2}(\hat{v}_2) \notin \hat{V}$. Indeed, we already excluded the case where $H_\rho(\hat{v}_2) = \hat{v}_2$, for every su-loop ρ. If ρ is an su-loop such that $H_\rho(\hat{v}_2) = \hat{v}_1$, then

$H_{\rho_1} H_\rho(\hat{v}_2) = H_{\rho_1}(\hat{v}_1) \notin \hat{V}$ by the initial assumption. Let ρ_2 be an su-loop such that $H_{\rho_2}(\hat{v}_2) \notin \hat{V}$.
Case 2.1: Assume that $H_{\rho_1}(\hat{v}_2) = \hat{v}_2$ and $H_{\rho_2}(\hat{v}_1) = \hat{v}_1$. Here, there exists $n \in \mathbb{N}$ such that the map $H_{\rho_1} A^n(q) H_{\rho_2}$ satisfies $H_{\rho_1} A^n(q) H_{\rho_2}(\hat{v}_i) \notin \hat{V}$, for $i = 1, 2$.
Case 2.2: Assume that $H_{\rho_1}(\hat{v}_2) = \hat{v}_2$ and $H_{\rho_2}(\hat{v}_1) = \hat{v}_2$. Exactly the same as Case 1 with ρ_2 and $\hat{v}_2$ instead of ρ_1 and $\hat{v}_1$.

In any of the above described cases, we obtain a contradiction with the assumption (22). This concludes the proof of item 4.

Using the invariance principle, Theorem 3.6, and item 3 of Proposition 3.4, we have that item 5 holds. For a proof of item 6 see [21, Proposition 5.1]. □

We say that a pinching γ-fiber bunched cocycle $A \in C^\gamma(\Sigma, SL_2(\mathbb{R}))$ is *quasi-twisting*, if there exists only one u-state in $\mathcal{M}_\mu(F_A)$.

Proposition 3.8 *Let $A \in C^\gamma(\Sigma, SL_2(\mathbb{R}))$ be a γ-fiber bunched cocycle with $L(A) > 0$. Then either A (or A^{-1}) is quasi-twisting or A is continuously conjugated to a diagonal cocycle.*

Outline of the Proof If $L(A) > 0$ and A is not weak twisting, there exists a measure $m \in \mathcal{M}_\mu(F_A)$ which is a su-state (see item 4 of Proposition 3.7). If m is non-ergodic, then, by item 4 of Proposition 3.5, A is conjugated to a diagonal. Otherwise, $m \in \{m^s, m^u\}$ and so either A or A^{-1} is quasi-twisting. If A is twisting, then in particular A is quasi-twisting. □

Example 3.6 (Pinching and Twisting for Locally Constant Cocycles) Consider a strongly irreducible, locally constant cocycle $A = (A_1, \ldots, A_\kappa) \in SL_2(\mathbb{R})^\kappa$ such that the semi-group generated by $A_1, \ldots, A_\kappa$ is unbounded. Then, by Furstenberg's criterion, $L(A) > 0$. An argument similar to the proof of item 4 Proposition 3.7 guarantees that A is pinching and twisting. In this case (20) gives only the cocycle matrix since the holonomies are identity matrices.

Remark 3.8 In the case of Schrödinger cocycles

$$A_E(x) = \begin{pmatrix} \varphi(x) - E & -1 \\ 1 & 0 \end{pmatrix}$$

with $E \in \mathbb{R}$ and $\varphi : \Sigma \to \mathbb{R}$ γ-Hölder (see Example 1.3) an extensive study of the set of energies E that have zero exponents is done in [4]. In particular, they prove that this set is discrete, and if it have some globally fiber bunched condition (that we will not define here) it is finite.

3.5 Continuity of the Lyapunov Exponent for Fiber Bunched Cocycles

It is very useful to think in pinching and twisting as the equivalent conditions to unboundedness and strong irreducibility in the Furstenberg's criterion. Using that interpretation it is natural to wonder if pinching and weak twisting γ-fiber bunched cocycles are continuity points of Lyapunov exponent in the C^γ topology. That is indeed the case. Consider A_k converging to A and let $m^k \in \mathcal{M}_\mu(F_{A_k})$ ergodic u-states realizing the Lyapunov exponent, i.e.,

$$L(A_k) = \int_{\Sigma\times\mathbb{P}^1} \log\|A_k(x)\,v\|\; d\,m^k(x,\hat{v}). \tag{23}$$

This is possible since if $L(A_k) > 0$ we may choose $m^k = m^u_{A_k}$ and in the case that $L(A_k) = 0$, by the invariance principle Theorem 3.6, any ergodic measure m^k works.

Up to a subsequence, we may assume that m^k converges, and so it must converge to some measure $m \in \mathcal{M}_\mu(F_A)$. It is possible to show that limit is also an u-state for A. But, by the pinching and weak twisting condition on A, there is only one u-state, namely m^u. So,

$$\begin{aligned} L(A_k) &= \int_{\Sigma\times\mathbb{P}^1} \log\|A_k(x)\,v\|\; d\,m^k(x,\hat{v}) \\ &\to \int_{\Sigma\times\mathbb{P}^1} \log\|A(x)\,v\|\; d\,m^u(x,\hat{v}) = L(A). \end{aligned} \tag{24}$$

The above discussion is summarized in the next proposition.

Proposition 3.9 *If $A \in C^\gamma(\Sigma, SL_2(\mathbb{R}))$ is a pinching and weak twisting γ-fiber bunched cocycle, then L is continuous at A in the C^γ-topology.*

Example 3.7 If A is an irreducible locally constant γ-fiber bunched cocycle, we claim that A is a continuity point of the Lyapunov exponent in the Hölder topology, $\gamma > 0$ (recall that the same is not true for $\gamma = 0$). Indeed, if $L(A) = 0$, there is nothing to prove. Assume that $L(A) > 0$, then by item 4 of Proposition 2.2 A is strongly irreducible. By Example 3.6, we see that A is pinching and twisting. The claim follows by Proposition 3.9.

More generally, assume that $A \in C^\gamma(\Sigma, SL_2(\mathbb{R}))$ is a γ-fiber bunched cocycle with $A(x) = A(x^+)$, for every $x \in \Sigma$. It was proved in [70, Theorem A] that if additionally A is pinching and weak twisting, then A is a continuity point of the Lyapunov exponent in the C^0-topology. A consequence of this result is that for non-invertible base dynamics, Bochi-Mane Theorem 1.2 is no longer true.

Now, assume that the fiber bunched cocycles A is not weak twisting and we investigate the continuity of the Lyapunov exponent at A.

Again, it is enough to consider the case $L(A) > 0$. In this case, using item 4 of Proposition 3.7, there exists a measure $\tilde{m} \in \mathcal{M}_{\mu}(F_A)$ which is a su-state. If $\tilde{m} \in \{m^s, m^u\}$ and this is the unique su-state, then either A or A^{-1} is quasi-twisting and a similar argument to prove continuity of L used in the weak twisting case above works. For that reason, we may assume that, up to continuous conjugacy, A is diagonal of the form (see Proposition 3.8)

$$A(x) = \begin{pmatrix} a(x) & 0 \\ 0 & a(x)^{-1} \end{pmatrix}.$$

Consider $A_k \to A$ and $m^k \to m$ as above. In particular, by the fact that A is diagonal, we have that m must to have the following disintegration

$$m_x = a\,\delta_{\hat{e}_1} + b\,\delta_{\hat{e}_2},$$

where $a, b \geq 0$, $a + b = 1$ and $\hat{e}_1, \hat{e}_2$ are the unitary vectors of the canonical basis of $\mathbb{R}^2$.

A technical issue that appears in the non locally constant case is that, even though the u-states for A_k, m^k, converges to an u-state, m, for A we are not able to guarantee that the disintegrations m^k_x converges to m_x for $x \in \Sigma$. That is mainly due to the lack of regularity of these disintegrations (which in general are only measurable). To overcome this issue we use the existence of stable holonomies (uniformly in a neighborhood of A) to change coordinates and restrict to the case where A_k and A only depends on the positive coordinates (see Proposition 3.3). This is very useful by the following reason: now we are able to project the measures m^k and m in the unilateral system $\Sigma^+ \times \mathbb{P}^1$ obtaining measures $m^{k,+}$ (ergodic) and m^+, invariant under $F^+_{A_k}$ and F^+_A respectively, such that

1. The disintegrations $\{m^{k,+}_{x^+}\}_{x^+ \in \Sigma^+}$ and $\{m^+_{x^+}\}_{x^+ \in \Sigma^+}$ are continuous disintegrations of $m^{k,+}$ and m^+;
2. For every $x^+ \in \Sigma^+$, $m^{k,+}_{x^+}$ converges to $m^+_{x^+}$ in the weak* topology.

Now, we proceed similarly to the locally constant case. Elaborating more on that, first we use a variation of the "energy argument" (using the fact that A is diagonal and $L(A) > 0$) to guarantee that the probabilities $m^{k,+}_{x^+}$ must to be atomic.

The proof follows by a classification of the atomic ergodic measures $m^{k,+}$ that can approach m. By item 3 of Proposition 3.7, there exists a periodic point $q \in \Sigma$, such that $A^l(q)$ is a hyperbolic matrix. So, $A^l_k(q)$ is also hyperbolic for every k sufficiently large (we assume that for every k). This, jointly with the continuity of the disintegration of $m^{k,+}$ in Σ^+, produces constraints for the structure of the measures $m^{k,+}$. In either case, the expression (24) is verified. This is summarized in the following result due to Backes, Butler and Brown.

Theorem 3.1 (Backes et al. [8]) *The map $FB^{\gamma}(\Sigma) \ni A \mapsto L(A)$ is continuous.*

This result shows that the problem of understanding the Lyapunov exponent function in the C^γ-topology, $\gamma > 0$, is completely different from the problem in the C^0-topology. Recall that in the C^0-topology, cocycles with non-zero Lyapunov exponent that are not uniformly hyperbolic (e.g., Example 3.1) are discontinuity points of L. Another important consequence of Theorem 3.1 is that now we have open sets of non-hyperbolic cocycles inside $C^\gamma(\Sigma, SL_2(\mathbb{R}))$ in which the Lyapunov exponent does not vanish.

Example 3.8 (Discontinuity Example) In Bocker-Viana example (Example 3.1), it is possible to give a set of parameters $1 \leq \alpha \leq \beta$ for which $A_{\alpha,\beta}$ is a discontinuity point of the Lyapunov exponent in the C^γ-topology. Indeed, a direct adaptation of the argument in [69, Theorem 9.22] shows that if $\alpha > \zeta^{2\gamma}$, then $A_{\alpha,\beta}$ is a discontinuity point of the Lyapunov exponent (recall that $A_{\alpha,\beta}$ is γ-fiber bunched for $\beta < \zeta^{\gamma/2}$).

Non-uniformly Fiber Bunched Cocycles As mentioned before, the existence of uniform holonomies in a neighborhood of the reference cocycle is the main ingredient in the proof of Theorem 3.1. So, it is important to understand more general conditions that guarantee the existence of holonomies in a stable way. A common approach in non-uniformly hyperbolic dynamics is to try to flexibilize a uniform concept such as γ-fiber bunched considering its asymptotic version: a cocycle A is called *non-uniformly γ fiber bunched* if "it has small Lyapunov exponent", more precisely, if A is γ-Hölder continuous and

$$L(A) < \frac{\gamma}{2} \log \zeta.$$

Recall that $\zeta > 1$ comes from the metric on Σ. The notion of non-uniformly γ-fiber bunched cocycles was first explored in [68] in which many of its properties were established (the term dominated was used instead of non-uniformly fiber bunched). We list a few of these properties in the next proposition.

Proposition 3.11 *It holds that,*

1. *If A is γ-fiber bunched, then A is non-uniformly γ-fiber bunched;*
2. *Non-uniformly γ-fiber bunched is an open condition in $C^\gamma(\Sigma, SL_2(\mathbb{R}))$;*
3. *If A is non-uniformly γ-fiber bunched, then A admits (non-uniformly) stable and unstable holonomies.*

Example 3.9 (Non-uniformly Fiber Bunched Cocycles) Let $A_{\alpha,\beta}$, $1 \leq \alpha \leq \beta$ be the cocycle given in the Bocker-Viana Example 3.1. Recall that,

$$L(A_{\alpha,\beta}) = |p_1 \log \beta - p_2 \log \alpha|.$$

So, $A_{\alpha,\beta}$ is non-uniformly γ-fiber bunched as long as $|p_1 \log \beta - p_2 \log \beta| < \frac{\gamma}{2} \log \zeta$. We could consider, for instance, a probability vector $p = (p_1, p_2)$ such that p_1/p_2 is very close to $\log \alpha / \log \beta$. So, this family provides examples of cocycles that are not fiber bunched (we could consider $\beta > \zeta^{\gamma/2}$), but still non-uniformly γ-fiber bunched.

The main goal in [68], which initiated the study of non-uniformly fiber bunched cocycles, was to obtain the next result.

Theorem 3.2 (Viana [68]) *The set of cocycles $A \in C^\gamma(\Sigma, SL_2(\mathbb{R}))$ with positive Lyapunov exponent is open and dense in $C^\gamma(\Sigma, SL_2(\mathbb{R}))$.*

The idea for proving this result is to start with a cocycle with zero Lyapunov exponent. Then it is automatically non-uniformly fiber bunched. With this condition non-uniform holonomies can be constructed in large measure sets. The difficult part is to construct this large measure set having product structure and periodic points. Once this is done using a version of the invariance principle on this set and a pinching and twisting condition the zero exponent can be destroyed.

Even though Theorem 3.2 guarantees that the set of zero Lyapunov exponent is rare inside of $C^\gamma(\Sigma, SL_2(\mathbb{R}))$, we still could have density of the set of cocycle with small Lyapunov exponent outside of the uniformly hyperbolic in $C^\gamma(\Sigma, SL_2(\mathbb{R}))$. That would be a similar phenomena to what happens in the case of C^0-topology and Theorem 1.2.

Problem 3.2 (Viana) *Are non-uniformly γ-fiber bunched cocycles dense in $C^\gamma(\Sigma, SL_2(\mathbb{R}))$ outside of the uniformly hyperbolic cocycles?*

The concept of non-uniform fiber bunched was further explored to extend the result of continuity of the Lyapunov exponent.

Theorem 3.3 (Freijo and Marin [36]) *Assume that A is non-uniformly γ-fiber bunched admitting a family of uniform stable holonomies. Then A is a continuity point of the Lyapunov exponent in the C^γ-topology.*

In the same spirit as in the Pesin's theory, non-uniformly fiber-bunched cocycles admits very large sets of the base space where the holonomies are indeed uniform. In this large probability sets, we may work as in the Theorem 3.1 since we have uniformity of the holonomies. The main issue is to have a control of the lack of uniformity in the small probability sets to obtain the properties 1 and 2 above. This is handled by a careful analysis of the convergence of the disintegrations $m^{k,+}$ and $m^{k,-}$ in these large probability sets.

A nice consequence of the techniques obtained in [36] is the following: if A is an irreducible non-uniformly γ-fiber bunched cocycle, then A is a continuity point of the Lyapunov exponent in the C^γ-topology. Indeed, assuming that $L(A) > 0$ and so A is strongly irreducible, by Example 3.6, we see that A is pinching and twisting. In particular, there is only one u-state, namely m^u. However, the following fact may be extended from the context of fiber bunched cocycles to the non-uniformly fiber bunched cocycles: if A_k is a sequence converging to A in the C^γ-topology and m^k is a sequence of u-states for A_k realizing $L(A_k)$ (Eq. (23)), then, up to a subsequence, the limit of m^k is still a u-state for A. This is enough to conclude the continuity.

For cocycles which are not non-uniformly fiber-bunched(large Lyapunov exponent) the problem of understanding the continuity properties of the Lyapunov exponent is still open. However, a few examples of discontinuity have been explored in the literature.

Example 3.10 (Butler [25]) In the Bocker-Viana's Example 3.1, assume that $\alpha = \beta > 1$ and write $A_\beta := A_{\alpha,\beta}$. For the probability vector $p = (p_1, p_2)$, assume that $p_1 \in (1/2, 1)$. Then, in [25], Butler showed that A_β is a discontinuity of the Lyapunov exponent if $\beta^{p_1 - p_2} \geq \zeta^{\gamma/2}$. Recall, from Example 3.9, that A_β is non-uniformly fiber bunched if $\beta^{p_1-p_2} < \zeta^{\gamma/2}$. So, as long as A_β is not non-uniformly γ-fiber bunched this cocycle is a discontinuity point of the Lyapunov exponent.

Remark 3.9 The arguments used [25] may also be used to guarantee that if $\beta > \zeta^\gamma$, then A_β is a discontinuity point of the Lyapunov exponent in the C^γ-topology. This is a slight improvement of the range presented in Example 3.1. Recall that in this case, the γ-fiber bunched condition is equivalent to $\beta < \zeta^{\gamma/2}$.

Remark 3.10 Observe that the Theorem 3.3 (or the comment that follows it) does not cover the case of A_β for any $\beta > 1$, even if we assume the non-uniform fiber bunched condition (A_β is reducible).

Problem 3.3 *Is A_β, defined above, a continuity point of the Lyapunov exponent in the C^γ-topology, for $\beta \in (\zeta^{\gamma/2}, \zeta^\gamma)$ and $\beta^{p_1-p_2} < \zeta^{\gamma/2}$?*

3.6 Modulus of Continuity of LE for Fiber Bunched Cocycles

Now we come back to the world of (uniform) fiber-bunched cocycles to discuss the modulus of continuity of the Lyapunov exponent. The machine developed in [32] to obtain Hölder regularity of the Lyapunov through large deviation estimates may also be applied to the context of cocycles $A : \Sigma \to SL_2(\mathbb{R})$ depending on infinitely many coordinates. That is the content of the next theorem.

Theorem 3.4 (Duarte et al. [33]) *Assume that the γ-fiber bunched cocycle A satisfies the pinching and twisting condition. Then, there exist $\theta > 0$ and a neighborhood $\mathcal{U} \subset C^\gamma(\Sigma, SL_2(\mathbb{R}))$ of A such that, restricted to this neighborhood, the Lyapunov exponent function L is θ-Hölder continuous.*

In order to obtain large deviation estimates in this context we follow the same lines of the proof of Theorem 2.6. Below, we mention some of the required adaptations.

To start, we use the families of unstable holonomies, which are uniform in a neighborhood of A where the fiber bunched condition holds, to reduce the analysis to maps depending only on negative coordinates (see Proposition 3.3) i.e., it is enough to assume that $A(x) = A(x^-)$, for every $x \in \Sigma$, and guarantee that there exist constants $\delta, C, \kappa, \varepsilon_0 > 0$ such that for every cocycle $B \in C^\gamma(\Sigma, SL_2(\mathbb{R}))$, $\|A - B\|_{C^\gamma} < \delta$, with $B(x) = B(x^-)$, for every $x \in \Sigma$, and for every $\varepsilon \in (0, \varepsilon_0)$

we have that

$$\mu\left(\left\{x \in \Sigma : \left|\frac{1}{n}\log\|B^n(x)\| - L(B)\right| > \varepsilon\right\}\right) \leq C\, e^{-\kappa\varepsilon^2 n}.$$

After this reduction, we proceed with the analysis of the spectral properties of the Markov operator, but differently of the Step 1 in Theorem 2.6, the cocycle A may depend on infinitely many coordinates, so there is a small adaptation on the definition of the Markov operator for this system. Indeed, set $Q : C^0(\Sigma^- \times \mathbb{P}^1) \to C^0(\Sigma^- \times \mathbb{P}^1)$ given by

$$Q_A(\varphi)(x^-, \hat{v}) := \sum_{i=1}^{\kappa} p_i\, \varphi((x^-, i), A(x^-)\,\hat{v}).$$

It is not hard to see that the measure m^u on $\Sigma^- \times \mathbb{P}^1$ is a fixed point of the adjoint action Q^* (this is the definition of stationary measure in this context).

The spectral analysis of Q_A is concluded once we notice that: First, defining

$$K_{n,\theta}(A) := \sup_{x\in\Sigma}\, \sup_{\hat{u}\neq\hat{v}} \int_\Sigma \left(\frac{d(A^n(x)\,\hat{u},\, A^n(x)\,\hat{v})}{d(\hat{u}, \hat{v})}\right)^\theta d\mu(x),$$

for any $\theta \in (0, 1)$ there exists $C > 0$ such that for every $\varphi \in C^\theta(\Sigma^- \times \mathbb{P}^1)$,

$$[Q_A^n(\varphi)]_\theta \leq K_{n,\theta}(A)\, [\varphi]_\theta + C\, \|\varphi\|_\infty\,.$$

Here, we are using the slightly different θ-Hölder norm,

$$[\varphi]_\theta := \sup_{x^-\in\Sigma^-} [\varphi(x^-, \cdot)]_\theta + \sup_{\hat{u}\in\mathbb{P}^1} [\varphi(\cdot, \hat{u})]_\theta\,.$$

Second, equivalently to the Step 1 of Theorem 2.6 (see Eq. (10) and the discussion that follows), the positivity of $L(A)$ and the pinching and twisting assumption is invoked to guarantee the exponential decay of the quantities $K_{n,\theta}(A)$ for some $\theta_0 \in (0, 1)$. So, using these considerations there exist constants $\lambda, C > 0$ such that for every $\varphi \in C^{\theta_0}(\Sigma^- \times \mathbb{P}^1)$ and every n sufficiently large,

$$[Q^n(\varphi)]_{\theta_0} \leq e^{-\lambda n}[\varphi]_{\theta_0} + C\, \|\varphi\|_\infty\,.$$

These ingredients suffice to guarantee the (uniform in a neighborhood of A) spectral gap of the operator Q_A. To conclude the regularity of the Lyapunov exponent, we follow the rest of the steps mentioned in the proof of Theorem 2.6.

Remark 3.11 Similar large deviation principle and limit theorems for this class of cocycles were obtained by Park and Piraino [57].

It is natural to wonder what happens if we no longer assume the pinching and twisting condition.

Problem 3.4 *What is the modulus of continuity of the Lyapunov exponent near a fiber bunched diagonal cocycle (see Example 3.1)?*

The solution of this problem would give an extension of the Theorem 2.8 from the locally constant cocycle to a the fiber bunched context.

The precise modulus of continuity of the Lyapunov exponent is still far from being established. However, we could consider a generalization of Theorem 2.11 to the context of γ-fiber bunched cocycles. To better pose the problem, we adapt the concept of heteroclinic tangency introduced in Sect. 2.4.

We say that a cocycle $A \in C^{\gamma}(\Sigma, SL_2(\mathbb{R}))$ admits a *heteroclinic tangency* if there exist periodic points $q_1, q_2 \in \Sigma$, of period m_1, m_2 respectively, and a heteroclinic point $z \in W^u_{\text{loc}}(q_1) \cap \sigma^{-k}(W^s_{\text{loc}}(q_2))$ such that

(i) $A^{m_i}(q_i)$ is hyperbolic for $i = 1, 2$;
(ii) For any $i = 1, 2$, let $\hat{v}_+(q_i), \hat{v}_-(q_i) \in \mathbb{P}^1$ be the eigen-directions $A^{m_i}(q_i)$ associated respectively with the largest and smallest (in absolute value) eigenvalues. Then,

$$H^s_{z,q_2}\, A^k(z)\, H^u_{q_1,z}\, \hat{v}_+(q_1) = \hat{v}_-(q_2).$$

Problem 3.5 *Let $A \in C^{\gamma}(\Sigma, SL_2(\mathbb{R}))$ be a pinching and twisting γ-fiber bunched cocycle admitting a heteroclinic tangency. In that context, by Theorem 3.4, we know that L is θ-Hölder continuous in a neighborhood of A and some $\theta > 0$. Does that imply that $\theta < H(\mu)/L(A)$?*

4 Conclusion and General Models

Positivity and continuity of the Lyapunov exponent in the classical setting of random products of matrices is well understood. The main elements in their study are the stationary measures and the structure of the group of matrices generated by the cocycle. Most of the techniques used in the locally constant case may be adapted, using holonomies, to deal with Hölder cocycles.

The strategy to study these cocycles uses holonomies to reduce the analysis, up to conjugacy, to cocycles only depending on the positive coordinates. These can be analyzed as cocycles over one-sided dynamics that, with proper adaptations, share many properties with the locally constant cocycles. The similarity is expressed for instance in the properties of invariant measures by the action of the cocycle in the one-sided product space. Indeed, u-states have in the general setting the central role played by the stationary measures in the locally constant case. These measures have a nice regularity which allowed us to recover many of the properties presented earlier for the stationary measures. However, for general cocycles, beyond the locally constant ones, the lack of independence in the matrix products along the orbit is an important issue which requires suitable adaptations. Concepts such as strong irreducibility or non compactness of the group generated by the matrices of the cocycle are reformulated in terms of periodic points and su-loops. These mentioned adaptations are only possible for cocycles admitting families of stable and unstable linear holonomies. Not much is known for more general cocycles.

Most of the results mentioned in the previous section can be stated in some more general context. We could consider on the base measures that are not necessarily Bernoulli, but have some "good" product structure which include Markov shifts. Nevertheless, because we did not want to focus on the structure of the measure but instead on the important properties of the fiber action, we preferred to state the results for Bernoulli measures which automatically satisfy all the hypothesis in the previous theorems.

Now to finish we recall some recent results and open problems on more general base dynamics with some hyperbolicity.

4.1 More General Dynamics

Recall that, by the classical construction of Markov partitions for Anosov maps, see for example [24], Anosov diffeomorphisms can be semi-conjugated to Markov shifts. Via this conjugacy the previous results can be extended to linear cocycles over Anosov diffeomorphisms.

So, what happens beyond uniform hyperbolic base maps?

Beyond uniform hyperbolicity, the two main classes of dynamics with some hyperbolicity that have been more studied are the *partially hyperbolic* and the *non-uniformly hyperbolic diffeomorphisms*.

There is some advance in the study of continuity and positivity of the exponents for cocycles over these maps.

For non-uniformly hyperbolic base dynamics with measures with product structure we have the following result:

- There exists an open and dense set of Hölder cocycles with positive Lyapunov exponent, see [68].

Problem 4.1 *Is the Lyapunov exponent continuous with in the space of Hölder fiber bunched cocycles over non uniformly hyperbolic maps?*

In this case, fiber bunched means that the inequality (18) is satisfied with e^{λ} instead of ζ, where λ is the minimum of the Lyapunov exponents of the base map's derivative.

By results in [62] and [11], non-uniformly hyperbolic base dynamics may be codified using a symbolic model such as the one presented in these notes, see [10] where this idea was used. The drawback is that such symbolic dynamics now could have countably infinitely many symbols. In this case, the problem of continuity of the Lyapunov exponent for more general non-uniformly hyperbolic maps may be approached from the symbolic point of view but now over non-compact spaces, see [65].

Below, we mention a few results in the context where the base dynamics is a partially hyperbolic map.

- Existence of an open and dense set of Hölder cocycles with positive Lyapunov exponent. For volume preserving accessible maps, see [6]. For skew products, see [59];
- Criterion for continuity of the Lyapunov spectrum for skew products, see [60];
- Openness of the set of fiber bunched Hölder cocycles with positive Lyapunov exponent. For volume preserving accessible maps, see [9].

Fiber bunched for partially hyperbolic maps means that the inequality (17) is satisfied with ζ equal to the minimum of the expansion of the diffeomorphism along the unstable direction and its inverse along the stable.

For partially hyperbolic maps, there are examples of Hölder cocycles with positive exponent accumulated by cocycles with zero exponent, even in the fiber bunched case, see [9]. These examples are constructed over a direct product of an Anosov with a rotation, in particular they are not accessible. For volume preserving accessible diffeomorphisms, this type of discontinuity can not happen, as stated in [9].

Problem 4.2 *Assume that the base dynamics is a partially hyperbolic volume preserving accessible diffeomorphism. Does the Lyapunov exponent varies continuously in the space of Hölder fiber bunched cocycles.*

Let us introduce now a class of partially hyperbolic systems that we call *mixed models*. These are maps of the form

$$f : \Sigma \times S^1 \to \Sigma \times S^1, \quad f(x, t) = (\sigma(x), t + \theta_x),$$

where S^1 is the unit circle and $\theta : \Sigma \to S^1$ is a continuous function. These maps preserve the measure $\mu \times dt$, where dt is the Lebesgue measure on S^1. In the same way that the full Bernoulli shift (Σ, σ, μ) provides a basic model for hyperbolic systems, mixed models are a basic class of partially hyperbolic systems with compact fibers where the action is elliptic.

Remark 4.1 Linear cocycles over torus rotations, known as quasi periodic cocycles, have been extensively studied. The lack of hyperbolicity in the base dynamics requires completely different techniques from the ones mentioned in these notes. For results regarding positivity and continuity of the Lyapunov exponent in this context, see [32], [30] (for the Schrödinger context [28]) and references therein.

An interesting class of linear cocycles over mixed models is the so called *mixed random quasi-periodic cocycles*, which corresponds to random products of quasi periodic cocycles. Consider a finite measure $\sum_{i=1}^{\kappa} p_i \delta_{(\theta_i, A_i)}$, where (θ_i, A_i)

defines a quasi-periodic cocycle. Analogously to the definition of random cocycles, we consider a "locally constant" mixed cocycle, defined by $A(x, t) := A_{x_0}(t)$, over the mixed base dynamics, $f(x, t) = (\sigma(x), t + \theta_{x_0})$.

Recently, this type of model has been attracted a lot of attention. We list a few results in this regard.

- Existence of open and dense set of quasi-periodic cocycles $(A_1, \ldots, A_\kappa)$ in which the Lyapunov exponent function is continuous and positive; see [15].
- A criterion for positivity of the exponents, see [26].
- Uniform upper semi-continuity of the exponents [27].
- Analiticity of the Lyapunov exponent with respect to the probability vector [16].

In the space of cocycles $A : \Sigma \times S^1 \to SL_2(\mathbb{R})$ depending only on x_0 and smoothly on $t \in S^1$, there are examples of discontinuity of the Lyapunov exponent function, see [9, section 5.2]. Here, the discontinuity comes from the discontinuity of smooth quasi periodic cocycles, see [71].

It is known that for quasi periodic analytic cocycles, the exponent varies continuously, see [5]. So, we have the following question:

Problem 4.3 *Is the Lyapunov exponent a continuous function in the space of analytic cocycles (with respect to the circle coordinate) among the mixed random quasi periodic cocycles?*

As mentioned in [15], there are open and dense conditions to guarantee continuity of the exponents (a generalization of the pinching and twisting condition). So, it is natural to ask the following:

Problem 4.4 *In the set of pinching and twisting cocycles (see the definition [15]) of the mixed random quasi periodic cocycles, does the exponent vary Hölder continuously?*

As we can see, there are many open questions to be explored about the behavior of the Lyapunov exponent function, especially beyond some restricted classes such as the random locally constant cocycles, the fiber bunched cocycles over hyperbolic maps and the quasi-periodic cocycles.

Acknowledgments The authors would like to express their gratitude to P. Duarte for the invitation to write this survey and for his valuable suggestions that helped to improve the quality of text. The authors would like to thank the organizers of the workshop "New trends in Lyapunov exponents", where discussions on the elaboration of this material began. The authors would also like to thank Graccyela Salcedo by the careful reading and suggestions on the final version of the text. The gratitude of the authors also extends to Łukasz Rzepnicki for providing Fig. 2.

Both authors were supported by FCT-Fundação para a Ciência e a Tecnologia through the project PTDC/MAT-PUR/29126/2017. J. B. was supported by the Center of Excellence "Dynamics, Mathematical Analysis and Artificial Intelligence" at Nicolaus Copernicus University in Toruń. M.P. was supported by CNPQ, Instituto Serrapilheira, grant "Jangada Dinâmica: Impulsionando Sistemas Dinâmicos na Região Nordeste" and the Coordenação de Aperfeiçoamento de Pessoal de Nível Superior—Brasil (CAPES)—Finance Code 001.

References

1. A. Avila. Density of positive lyapunov exponents for sl(2, r)-cocycles. *Journal of the American Mathematical Society*, 24(4):999–1014, 2011.
2. A. Avila and J. Bochi. A formula with some applications to the theory of lyapunov exponents. *Israel Journal of Mathematics*, 131(1):125–137, 2002.
3. A. Avila, J. Bochi, and J.-C. Yoccoz. Uniformly hyperbolic finite-valued sl (2, r)-cocycles. *Commentarii Mathematici Helvetici*, 85(4):813–884, 2010.
4. A. Avila, D. Damanik, and Z. Zhang. Schrödinger operators with potentials generated by hyperbolic transformations: I—positivity of the lyapunov exponent. *Inventiones mathematicae*, 231(2):851–927, 2023.
5. A. Avila, S. Jitomirskaya, and C. Sadel. Complex one-frequency cocycles. *Journal of the European Mathematical Society*, 16(9):1915–1935, 2014.
6. A. Avila, J. Santamaria, and M. Viana. Holonomy invariance: rough regularity and applications to Lyapunov exponents. *Astérisque*, 358:13–74, 2013.
7. A. Avila and M. Viana. Extremal Lyapunov exponents: an invariance principle and applications. *Invent. Math.*, 181(1):115–189, 2010.
8. L. Backes, A. Brown, and C. Butler. Continuity of Lyapunov exponents for cocycles with invariant holonomies. *J. Mod. Dyn.*, 12:223–260, 2018.
9. L. Backes, M. Poletti, and A. Sánchez. The set of fiber-bunched cocyles with nonvanishing lyapunov exponents over a partially hyperbolic map is open. *Mathematical Research Letters*, 25, 2018.
10. L. Backes, M. Poletti, P. Varandas, and Y. Lima. Simplicity of lyapunov spectrum for linear cocycles over non-uniformly hyperbolic systems. *Ergodic Theory and Dynamical Systems*, page 1–23, 2019.
11. S. Ben Ovadia. Symbolic dynamics for non-uniformly hyperbolic diffeomorphisms of compact smooth manifolds. *Journal of Modern Dynamics*, 13:43–113, 2018.
12. Y. Benoist and J.-F. Quint. On the regularity of stationary measures. *Israel Journal of Mathematics*, 226:1–14, 2018.
13. J. Bezerra, A. Cai, P. Duarte, C. Freijo, and S. Klein. A dynamical thouless formula. *arXiv preprint arXiv:2208.06022*, 2022.
14. J. Bezerra and P. Duarte. Upper bound on the regularity of the lyapunov exponent for random products of matrices. *arXiv preprint arXiv:2208.03575*, 2022.
15. J. Bezerra and M. Poletti. Random product of quasi-periodic cocycles. *Proceedings of the AMS*, 2021.
16. J. Bezerra, A. Sánchez, and E. H. Y. Tall. Analyticity of the lyapunov exponents of random products of quasi-periodic cocycles. preprint arxiv:2111.00683.
17. J. Bochi. Genericity of zero lyapunov exponents. *Ergodic Theory and Dynamical Systems*, 22(6):1667–1696, 2002.
18. C. Bocker-Neto and M. Viana. Continuity of lyapunov exponents for random two-dimensional matrices. *Ergodic Theory and Dynamical Systems*, 37(5):1413–1442, 2017.
19. C. Bonatti, X. Gómez-Mont, and M. Viana. Généricité d'exposants de lyapunov non-nuls pour des produits déterministes de matrices. *Annales de l'Institut Henri Poincaré C*, 20(4):579–624, 2003.
20. C. Bonatti and M. Viana. Lyapunov exponents with multiplicity 1 for deterministic products of matrices. *Ergod. Th. & Dynam. Sys*, 24:1295–1330, 2004.
21. C. Bonatti and M. Viana. Lyapunov exponents with multiplicity 1 for deterministic products of matrices. *Ergodic Theory and Dynamical Systems*, 24(5):1295–1330, 2004.
22. P. Bougerol and J. Lacroix. *Products of random matrices with applications to Schrödinger operators*, volume 8. Springer Science & Business Media, 2012.
23. J. Bourgain. Finitely supported measures on sl2(r) which are absolutely continuous at infinity, geometric aspect of functional analysis (b. klartag et al. eds.), 133–141. *Lecture Notes in Math*, 2050, 2012.

24. R. Bowen. *Equilibrium states and the ergodic theory of Anosov diffeomorphisms*, volume 470 of *Lect. Notes in Math.* Springer Verlag, 1975.
25. C. Butler. Discontinuity of lyapunov exponents near fiber bunched cocycles. *Ergodic Theory and Dynamical Systems*, 38(2):523–539, 2018.
26. A. Cai, P. Duarte, and S. Klein. Furstenberg theory of mixed random-quasiperiodic cocycles. preprint arxiv:2201.04745.
27. A. Cai, P. Duarte, and S. Klein. Mixed random-quasiperiodic cocycles. *Bulletin of the Brazilian Mathematical Society, New Series*, 53(4):1469–1497, 2022.
28. D. Damanik. Schrödinger operators with dynamically defined potentials. *Ergodic Theory and Dynamical Systems*, 37(6):1681–1764, 2017.
29. B. Deroin and R. Dujardin. Random walks, kleinian groups, and bifurcation currents. *Inventiones mathematicae*, 190(1):57–118, 2012.
30. P. Duarte and S. Klein. Continuity of the lyapunov exponents of linear cocycles. *Publicacoes Matematicas do IMPA, 31o Coloquio Brasileiro de Matematica IMPA*, 2017.
31. P. Duarte and S. Klein. Large deviations for products of random two dimensional matrices. *Communications in Mathematical Physics*, 375:2191–2257, 2020.
32. P. Duarte, S. Klein, et al. Lyapunov exponents of linear cocycles. *Atlantis Studies in Dynamical Systems*, 3, 2016.
33. P. Duarte, S. Klein, and M. Poletti. Hölder continuity of the lyapunov exponents of linear cocycles over hyperbolic maps. *Mathematische Zeitschrift*, 302(4):2285–2325, 2022.
34. P. Duarte, S. Klein, and M. Santos. A random cocycle with non hölder lyapunov exponent. *arXiv preprint arXiv:1811.02647*, 2018.
35. P. Erdös. On the smoothness properties of a family of bernoulli convolutions. *American Journal of Mathematics*, 62(1):180–186, 1940.
36. C. Freijo and K. Marin. Continuity of lyapunov exponents for non-uniformly fiber-bunched cocycles. *Ergodic Theory and Dynamical Systems*, 41(12):3740–3767, 2021.
37. A. Furman. Random walks on groups and random transformations. In *Handbook of dynamical systems*, volume 1, pages 931–1014. Elsevier, 2002.
38. H. Furstenberg and H. Kesten. Products of random matrices. *The Annals of Mathematical Statistics*, 31(2):457–469, 1960.
39. H. Furstenberg and Y. Kifer. Random matrix products and measures on projective spaces. *Israel Journal of Mathematics*, 46:12–32, 1983.
40. M. Goldstein and W. Schlag. Hölder continuity of the integrated density of states for quasi-periodic schrödinger equations and averages of shifts of subharmonic functions. *Annals of Mathematics*, pages 155–203, 2001.
41. A. Gorodetski and V. Kleptsyn. Parametric furstenberg theorem on random products of sl (2, r) matrices. *Advances in Mathematics*, 378:107522, 2021.
42. N. Gourmelon. Rotation numbers of perturbations of smooth dynamics. *arXiv preprint arXiv:2002.06783*, 2020.
43. Y. Guivarc'h and A. Raugi. Products of random matrices: convergence theorems. *Contemp. Math*, 50(31-54):2, 1986.
44. M. Hochman and B. Solomyak. On the dimension of furstenberg measure for sl _ 2 (r) sl 2 (r) random matrix products. *Inventiones mathematicae*, 210:815–875, 2017.
45. R. A. Johnson. Exponential dichotomy, rotation number, and linear differential operators with bounded coefficients. *Journal of differential equations*, 61(1):54–78, 1986.
46. B. Kalinin. Livšic theorem for matrix cocycles. *Annals of mathematics*, pages 1025–1042, 2011.
47. B. Kalinin and V. Sadovskaya. Holonomies and cohomology for cocycles over partially hyperbolic diffeomorphisms. *arXiv preprint arXiv:1408.5602*, 2014.
48. A. Katok and V. Niţică. *Rigidity in higher rank abelian group actions: Volume 1, Introduction and Cocycle Problem*, volume 185. Cambridge University Press, 2011.
49. Y. Kifer. Perturbations of random matrix products. *Zeitschrift für Wahrscheinlichkeitstheorie und Verwandte Gebiete*, 61(1):83–95, 1982.

50. Y. Kifer. *Ergodic theory of random transformations*, volume 10. Springer Science & Business Media, 2012.
51. F. Ledrappier. Positivity of the exponent for stationary sequences of matrices. In *Lyapunov Exponents: Proceedings of a Workshop held in Bremen, November 12–15, 1984*, pages 56–73. Springer, 2006.
52. F. Ledrappier and L.-S. Young. The metric entropy of diffeomorphisms: Part i: Characterization of measures satisfying pesin's entropy formula. *Annals of Mathematics*, pages 509–539, 1985.
53. F. Ledrappier and L.-S. Young. The metric entropy of diffeomorphisms: part ii: relations between entropy, exponents and dimension. *Annals of Mathematics*, pages 540–574, 1985.
54. P. Mattila. *Fourier analysis and Hausdorff dimension*, volume 150. Cambridge University Press, 2015.
55. V. I. Oseledec. A multiplicative ergodic theorem, lyapunov characteristic numbers for dynamical systems. *Transactions of the Moscow Mathematical Society*, 19:197–231, 1968.
56. É. l. Page. Régularité du plus grand exposant caractéristique des produits de matrices aléatoires indépendantes et applications. In *Annales de l'IHP Probabilités et statistiques*, volume 25, pages 109–142, 1989.
57. K. Park and M. Piraino. Transfer operators and limit laws for typical cocycles. *Communications in Mathematical Physics*, 389(3):1475–1523, 2022.
58. Y. Peres. Analytic dependence of lyapunov exponents on transition probabilities. In *Lyapunov Exponents: Proceedings of a Conference held in Oberwolfach, May 28–June 2, 1990*, pages 64–80. Springer, 2006.
59. M. Poletti. Stably positive lyapunov exponents for symplectic linear cocycles over partially hyperbolic diffeomorphisms. *Discrete and Continuous Dynamical Systems - A*, 38:5163, 2018.
60. M. Poletti and M. Viana. Simple lyapunov spectrum for certain linear cocycles over partially hyperbolic maps. *Nonlinearity*, 32(1):238, 2018.
61. D. Ruelle. Analycity properties of the characteristic exponents of random matrix products. *Advances in mathematics*, 32(1):68–80, 1979.
62. O. Sarig. Symbolic dynamics for surface diffeomorphisms with positive entropy. *J. Amer. Math. Soc.*, 26:341–426, 2013.
63. B. Simon and M. Taylor. Harmonic analysis on sl (2, r) and smoothness of the density of states in the one-dimensional anderson model. *Communications in mathematical physics*, 101:1–19, 1985.
64. B. Solomyak. On the random series $\sum \pm\lambda^n$ (an erdos problem). *Annals of Mathematics*, pages 611–625, 1995.
65. A. Sánchez and M. Viana. Lyapunov exponents of probability distributions with non-compact support. https://arxiv.org/abs/1810.03061.
66. E. H. Tall and M. Viana. Moduli of continuity for the lyapunov exponents of random sl(2)-cocycles. *Transactions of the American Mathematical Society*, 373(2):1343–1383, 2020.
67. D. J. Thouless. A relation between the density of states and range of localization for one dimensional random systems. *Journal of Physics C: Solid State Physics*, 5(1):77, 1972.
68. M. Viana. Almost all cocycles over any hyperbolic system have nonvanishing Lyapunov exponents. *Ann. of Math.*, 167:643–680, 2008.
69. M. Viana. *Lectures on Lyapunov exponents*, volume 145. Cambridge University Press, 2014.
70. M. Viana and J. Yang. Continuity of lyapunov exponents in the c 0 topology. *Israel Journal of Mathematics*, 229(1):461–485, 2019.
71. Y. Wang and J. You. Examples of discontinuity of lyapunov exponent in smooth quasiperiodic cocycles. *Duke Math. J.*, 162(13):2363–2412, 10 2013.

Randomness Versus Quasi-Periodicity

Ao Cai

Abstract This paper serves as an extended road map for our long-term project "Mixed Random-Quasiperiodic Cocycles" with Pedro Duarte and Silvius Klein. Despite exhibiting totally different natures, the world of linear random cocycles and that of linear quasi-periodic cocycles may still have potential relations that we intend to investigate. This was inspired by Jiangong You's intriguing question posed in 2018 on the stability of the Lyapunov exponent of quasi-periodic Schrödinger operators under random noise.

1 Introduction

The story began during an international conference on dynamical systems held at Nanjing University in 2018 when I met Pedro Duarte for the first time and Silvius Klein for the second time, who both became my postdoctor supervisors afterwards . Back then, I had just finished my PhD with my dissertation "Reducibility of finitely differentiable quasi-periodic cocycles and its applications" based on Kolmogorov-Arnold-Moser (KAM) theory under the supervision of Jiangong You. Obviously, I was a purely quasi-periodic person [1, 2, 8, 9] while Duarte and Klein had already many collaborations on random cocycles as well as quasi-periodic ones, see the two excellent books [10, 11] and the references therein.

Definitely, You's beautiful question motivated the two parallel rays to intersect and resonate intensively.

Digressions aside, You's original question states: "what is the behavior of the Lyapunov exponent of the quasi-periodic Schrödinger operator (cocycle) perturbed by an i.i.d. random noise?"

A. Cai (✉)
School of Mathematical Sciences, Soochow University, Suzhou, China
e-mail: acai@suda.edu.cn

© The Author(s), under exclusive license to Springer Nature Switzerland AG 2023
J. L. Dias et al. (eds.), *New Trends in Lyapunov Exponents*, CIM Series in Mathematical Sciences, https://doi.org/10.1007/978-3-031-41316-2_3

More precisely, consider the one dimensional discrete Schrödinger operator defined on $\ell^2(\mathbb{Z})$:

$$(H_{W,V,\alpha,\theta}u)_n = u_{n+1} + u_{n-1} + [V(\theta + n\alpha) + W_n]u_n$$

where V is a real-valued potential function and $\{W_n\}_{n\in\mathbb{Z}}$ is an i.i.d. sequence driven by some probability measure μ on $\mathbb{R}$. If we put a coupling multiplicative constant $\epsilon > 0$ before W_n and let ϵ go to 0, what will be the behavior of the Lyapunov exponent as a function of ϵ and μ? Namely, You's question is concerned with the stability of the Lyapunov exponent of quasi-periodic Schrödinger cocycles under random perturbation when the magnitude of the randomness vanishes.

This great question motivated us to use our entire toolbox to develop various new methods, trying to build the bridge between the quasi-periodicity and the randomness. For the time being, we still cannot give a complete answer about it but partial results are already available. In fact, along the whole way of attempting to answer it, we already acquired many fruitful results.

In this short paper, we provide five sections of results contained respectively in [3–7], where relations and comparisons of our theorems with the existing literature can be found. Since our main purpose is to convey ideas and state results with no proofs at all, we have decided to leave the statements of main theorems in their corresponding sections in order to avoid repetition.

This paper is organized as follows. In Sect. 2 we introduce the concept of mixed random-quasiperiodic cocycles. In Sect. 3, we state our Furstenberg positivity criterion of the maximal Lyapunov exponent of mixed models and give some applications in Mathematical Physics. Section 4 is an interlude which mainly contains an abstract large deviations type theorem. In Sect. 5 we talk about the Hölder continuity of the Lyapunov exponent. Finally, Sect. 6 is devoted to the stability of the Lyapunov exponent, the original motivating question.

2 Mixed Random-Quasiperiodic Cocycles

As preliminaries, in this section we introduce the concept of mixed random-quasiperiodic cocycles. We include the basic definitions but leave out most other details. Interested readers are kindly invited to read [4] for precise statements and proofs.

The Base Dynamics Let $(\Omega, \mathcal{B})$ be a standard Borel space and let $\nu \in \mathrm{Prob}_c(\Omega)$ be a compactly supported Borel probability measure on Ω. Regarding (Ω, ν) as a space of symbols, we consider the corresponding (invertible) Bernoulli system $(X, \sigma, \nu^{\mathbb{Z}})$, where $X := \Omega^{\mathbb{Z}}$ and $\sigma : X \to X$ is the (invertible) Bernoulli shift: for $\omega = \{\omega_n\}_{n\in\mathbb{Z}} \in X$, $\sigma\omega := \{\omega_{n+1}\}_{n\in\mathbb{Z}}$. Consider also its non invertible factor on $X^+ := \Omega^{\mathbb{N}}$.

Let $\mathbb{T}^d = (\mathbb{R}/\mathbb{Z})^d$ be the torus of dimension d, and denote by m the Haar measure on its Borel σ-algebra.

Given a continuous function $\mathfrak{a}: \Omega \to \mathbb{T}^d$, the skew-product map

$$f: X \times \mathbb{T}^d \to X \times \mathbb{T}^d, \quad f(\omega, \theta) := (\sigma\omega, \theta + \mathfrak{a}(\omega_0)) \tag{2.1}$$

will be referred to as a mixed random-quasiperiodic (base) dynamics.

This map preserves the measure $\nu^{\mathbb{Z}} \times m$ and it is the natural extension of the non-invertible map on $X^+ \times \mathbb{T}^d$ which preserves the measure $\nu^{\mathbb{N}} \times m$ and is defined by the same expression.

We call the measure ν ergodic, or ergodic with respect to f when the mixed random-quasiperiodic system $\left(X \times \mathbb{T}^d, f, \nu^{\mathbb{Z}} \times m\right)$ is ergodic. See [4, Section 2] for various characterizations of the ergodicity. In the same paper, a uniform convergence of Birkhoff sums of continuous observables to their space averages was established [4, Lemma 2.5]. This was further used, along with a stopping time argument, to prove a uniform base LDT theorem for continuous observables depending on finitely many coordinates [4, Theorem 2.4].

For our interest, before introducing the fiber dynamics, we are going to specify the Ω as follows.

The Group of Quasiperiodic Cocycles Given a frequency $\alpha \in \mathbb{T}^d$, let $\tau_\alpha(\theta) = \theta + \alpha$ be the corresponding ergodic translation on $\mathbb{T}^d$. Consider $A \in C^0(\mathbb{T}^d, \mathrm{SL}_m(\mathbb{R}))$ a continuous matrix valued function on the torus. A quasiperiodic cocycle is a skew-product map of the form

$$\mathbb{T}^d \times \mathbb{R}^m \ni (\theta, v) \mapsto (\tau_\alpha(\theta), A(\theta)v) \in \mathbb{T}^d \times \mathbb{R}^m .$$

This cocycle can thus be identified with the pair (α, A). Consider the set

$$\mathcal{G} = \mathcal{G}(d, m) := \mathbb{T}^d \times C^0(\mathbb{T}^d, \mathrm{SL}_m(\mathbb{R}))$$

of all quasiperiodic cocycles.

This set is a Polish metric space when equipped with the product metric (in the second component we consider the uniform distance). The space $\mathcal{G}$ is also a group, and in fact a topological group, with the natural composition and inversion operations

$$(\alpha, A) \circ (\beta, B) := (\alpha + \beta, (A \circ \tau_\beta)\, B)$$
$$(\alpha, A)^{-1} := (-\alpha, (A \circ \tau_{-\alpha})^{-1}) .$$

Given $\nu \in \mathrm{Prob}_c(\mathcal{G})$ let $\omega = \{\omega_n\}_{n \in \mathbb{Z}}$, $\omega_n = (\alpha_n, A_n)$ be an i.i.d. sequence of random variables in $\mathcal{G}$ with law ν. Consider the corresponding multiplicative process in the group $\mathcal{G}$

$$\Pi_n = \omega_{n-1} \circ \cdots \circ \omega_1 \circ \omega_0$$
$$= \left(\alpha_{n-1} + \cdots + \alpha_1 + \alpha_0,\ (A_{n-1} \circ \tau_{\alpha_{n-2}+\cdots+\alpha_0}) \cdots (A_1 \circ \tau_{\alpha_0})\, A_0\right).$$

In order to study this process in the framework of ergodic theory, we model it by the iterates of a linear cocycle.

The Fiber Dynamics Given $\nu \in \mathrm{Prob}_c(\mathcal{G})$, let $\Omega \subset \mathcal{G}$ be a closed subset such that $\Omega \supset \mathrm{supp}\nu$. We regard (Ω, ν) as a space of symbols and consider the shift σ on the space $X := \Omega^{\mathbb{Z}}$ of sequences $\omega = \{\omega_n\}_{n\in\mathbb{Z}}$ endowed with the product measure ν^Z and the product topology (which is metrizable). The standard projections

$$\mathfrak{a}\colon \Omega \to \mathbb{T}^d, \qquad \mathfrak{a}(\alpha, A) = \alpha$$
$$\mathcal{A}\colon \Omega \to C^0(\mathbb{T}^d, \mathrm{SL}_m(\mathbb{R})), \qquad \mathcal{A}(\alpha, A) = A$$

determine the linear cocycle $F = F_{(\mathfrak{a},\mathcal{A})}\colon X \times \mathbb{T}^d \times \mathbb{R}^m \to X \times \mathbb{T}^d \times \mathbb{R}^m$ defined by

$$F(\omega, \theta, v) := (\sigma\omega, \theta + \mathfrak{a}(\omega_0), \mathcal{A}(\omega_0)(\theta)\, v).$$

The non-invertible version of this map is defined similarly on $X^+ \times \mathbb{T}^d \times \mathbb{R}^m$, where $X^+ = \Omega^{\mathbb{N}}$.

Thus the base dynamics of the cocycle F is the mixed random-quasiperiodic map f defined above,

$$X \times \mathbb{T}^d \ni (\omega, \theta) \mapsto f(\omega, \theta) := (\sigma\omega, \theta + \mathfrak{a}(\omega_0)) \in X \times \mathbb{T}^d,$$

while the fiber action is induced by the map

$$X \times \mathbb{T}^d \ni (\omega, \theta) \mapsto \mathcal{A}(\omega, \theta) := \mathcal{A}(\omega_0)(\theta) \in \mathrm{SL}_m(\mathbb{R}).$$

The skew-product F will then be referred to as a *mixed random-quasiperiodic cocycle*. The space of mixed cocycles $F = F_{(\mathfrak{a},\mathcal{A})}$ is a metric space with the uniform distance

$$\mathrm{dist}\left((\mathfrak{a}, \mathcal{A}), (\mathfrak{a}', \mathcal{A}')\right) = \left\|\mathfrak{a} - \mathfrak{a}'\right\|_0 + \left\|\mathcal{A} - \mathcal{A}'\right\|_0.$$

For $\omega = \{\omega_n\}_{n\in\mathbb{Z}} \in X$ and $j \in \mathbb{N}$ consider the composition of random translations

$$\tau_\omega^j := \tau_{\mathfrak{a}(\omega_{j-1})} \circ \cdots \circ \tau_{\mathfrak{a}(\omega_0)} = \tau_{\mathfrak{a}(\omega_{j-1})+\cdots+\mathfrak{a}(\omega_0)} = \tau_{\mathfrak{a}(\omega_{j-1}\circ\cdots\circ\omega_0)}.$$

The iterates of the cocycle F are then given by

$$F^n(\omega, \theta, v) = \left(\sigma^n\omega, \tau_\omega^n(\theta), \mathcal{A}^n(\omega)(\theta)v\right),$$

where

$$\mathcal{A}^n(\omega) = \mathcal{A}(\omega_{n-1} \circ \cdots \circ \omega_1 \circ \omega_0)$$
$$= \left(\mathcal{A}(\omega_{n-1}) \circ \tau_\omega^{n-2}\right) \cdots \left(\mathcal{A}(\omega_1) \circ \tau_\omega^0\right) \mathcal{A}(\omega_0) .$$

Thus $\mathcal{A}^n(\omega)$ can be interpreted as a random product of quasiperiodic cocycles, driven by the measure ν on the group $\mathcal{G}$ of such cocycles. For convenience we also denote $\mathcal{A}^n(\omega, \theta) := \mathcal{A}^n(\omega)(\theta)$.

By Kingman's subadditive ergodic theorem, the limit of the sequence $\frac{1}{n} \log \|\mathcal{A}^n(\omega)(\theta)\|$ as $n \to \infty$ exists for $\nu^{\mathbb{Z}} \times m$ a.e. $(\omega, \theta) \in X \times \mathbb{T}^d$. Recall that for simplicity we call the measure ν ergodic if the base dynamics f is ergodic w.r.t. $\nu^{\mathbb{Z}} \times m$. In this case, the limit is a constant depending only on the measure ν and it is called the first (or maximal) Lyapunov exponent of the cocycle F, which we denote by $L_1(\nu)$.

Using the available uniform base LDT theorem as well as the subadditive ergodic theorem, we proved a uniform fiber upper LDT estimates [4, Theorem 3.1] which readily implies the upper semi-continuity of the Lyapunov exponent with respect to the Wasserstein distance. Moreover, at this level of generality, the uniform fiber lower LDT can not necessarily hold [4, Remark 3.2]. Otherwise we get continuity for free but there are counter examples cooked up based on Wang-You [15].

After the basic concepts and properties are introduced, we may start building various results upon them.

3 Positivity of the Lyapunov Exponent

Furstenberg Positivity Criterion The first interesting result is the following criterion for the positivity of the maximal Lyapunov exponent of mixed random-quasiperiodic cocycles.

Before stating the theorem, we introduce two concepts.

Definition 3.1 (See Definition 3.16 in [12]) Let $\mathcal{G}_0 \subset \mathcal{G}$ be a closed subgroup. We say that $\mathcal{G}_0$ is respectively *non-compact*, *strongly irreducible* or *Zariski dense* when there exists no measurable function $M \colon \mathbb{T}^d \to \mathrm{SL}_m(\mathbb{R})$ such that $C : \mathcal{G}_0 \times \mathbb{T}^d \to \mathrm{SL}_m(\mathbb{R})$,

$$C((\beta, B), \theta) := M(\theta + \beta)^{-1} B(\theta) M(\theta)$$

for any $(\beta, B) \in \mathcal{G}_0$ and m-a.e. θ, takes values in a proper closed subgroup $G_0 \subseteq \mathrm{SL}_m(\mathbb{R})$ that is compact, virtually reducible or algebraic, respectively.

Given measure $\nu \in \text{Prob}_c(\mathcal{G})$ we denote by $\mathcal{G}_\nu$ the closed subgroup of $\mathcal{G}$ generated by $\text{supp}(\nu)$. This measure ν is called *non-compact, strongly irreducible* or *Zariski dense* when the closed subgroup $\mathcal{G}_\mu$ satisfies the same properties.

Theorem 3.1 *Let $\nu \in \text{Prob}_c(\mathcal{G})$ and assume that ν is ergodic, non-compact and strongly irreducible. Then $L_1(\nu) > 0$.*

This naturally extends the classical Furstenberg's positivity criterion for products of random matrices [13] and its proof is indeed in the spirit of Furstenberg's original one. Moreover, it also has interesting applications in Mathematical Physics.

Applications to Schrödinger Operators Consider the Schrödinger operator $H(\theta)$ on $l^2(\mathbb{Z})$ defined by

$$(H(\theta)\psi)_n = -\psi_{n+1} - \psi_{n-1} + (v(\theta + n\alpha) + w_n)\ \psi_n \quad \forall n \in \mathbb{Z}, \tag{3.1}$$

for all $\psi = \{\psi_n\}_{n\in\mathbb{Z}} \in l^2(\mathbb{Z})$.

The i.i.d. sequence $\{w_n\}$ can be interpreted as a random perturbation of the quasiperiodic potential $\{v_n(\theta)\}$, where $v_n(\theta) := v(\theta + n\alpha)$.

We may instead randomize the frequency, that is, given an i.i.d sequence $\{\alpha_n\}_{n\in\mathbb{Z}}$ of random variables with values on the torus $\mathbb{T}^d$, consider the discrete Schrödinger operator

$$(H(\theta)\psi)_n = -\psi_{n+1} - \psi_{n-1} + v(\theta + \alpha_0 + \cdots + \alpha_{n-1})\ \psi_n \quad \forall n \in \mathbb{Z}. \tag{3.2}$$

We may certainly both randomize the frequency and randomly perturb the resulting potential (independently from each other or not), thus obtaining the mixed Schrödinger operator

$$(H(\theta)\psi)_n = -\psi_{n+1} - \psi_{n-1} + (v_n(\theta) + w_n)\ \psi_n \quad \forall n \in \mathbb{Z}, \tag{3.3}$$

where

$$v_n(\theta) = v(\theta + \alpha_0 + \cdots + \alpha_{n-1}) \quad \text{if} \quad n \geq 1.$$

For these three models, we have respectively three applications.

Define $P : \omega \mapsto \begin{pmatrix} 1 & \omega \\ 0 & 1 \end{pmatrix}$, $\mathbb{R} \to \text{SL}_2(\mathbb{R})$.

Denote by $S_E = \begin{pmatrix} V - E & -1 \\ 1 & 0 \end{pmatrix}$ the Schrödinger cocycle with energy E and potential V.

Proposition 3.1 *Consider the Schrödinger operator (3.1) with randomly perturbed quasiperiodic potential and the corresponding cocycles driven by the measures $\nu_E := \delta_\alpha \times \int_\mathbb{R} \delta_{P(\omega)S_E}\, d\rho(\omega)$ where $\rho \in \text{Prob}_c(\mathbb{R})$.*

If $\mathrm{supp}(\rho)$ *has more than one element, then* $L_1(\nu_E) > 0\ \forall E \in \mathbb{R}$.

We now consider the case of randomly chosen frequencies.

Proposition 3.2 *Consider the Schrödinger operator* (3.2) *with randomly perturbed frequencies and the corresponding cocycles driven by the measures* $\nu_E := \mu \times \delta_{S_E}$ *where* $\mu \in \mathrm{Prob}(\mathbb{T}^d)$.

If there are two frequencies $\alpha, \beta \in \mathrm{supp}(\mu)$ *such that* $\beta - \alpha$ *is an ergodic translation on* $\mathbb{T}^d$, *and if the potential function* $v(\theta)$ *is analytic and non-constant (or, more generally, if the continuous functions* $v(\theta)$ *and* $v(\theta + \beta - \alpha)$ *are transversal), then* $L_1(\nu_E) > 0\ \forall E \in \mathbb{R}$.

Proposition 3.3 *Consider the Schrödinger operator* (3.3) *with randomly perturbed frequencies and quasiperiodic potential and the corresponding cocycles driven by the measures* $\nu_E = \int_{\mathbb{T}^d \times \mathbb{R}} \delta_{(\alpha,\, P(\omega) S_E)}\, d\eta(\alpha, \omega)$ *where* $\eta \in \mathrm{Prob}_c(\mathbb{T}^d \times \mathbb{R})$.

If $\mathrm{supp}(\eta)$ *contains two points* (α, ω_1) *and* (α, ω_2) *with* α *rationally independent and* $\omega_1 \neq \omega_2$, *then* $L_1(\nu_E) > 0\ \forall E \in \mathbb{R}$.

It is not hard to see that randomness dominates quasi-periodicity in the sense of positive Lyapunov exponents. For example, any quasi-periodic Schrödinger operator in the almost reducible regime (small coupling constant, Diophantine frequency and smooth enough potential) has purely absolutely continuous spectrum with zero Lyapunov exponents [1, 2]. However, when the randomness comes in, the new mixed cocycle has positive Lyapunov exponent which is consistent with the random case. Indeed, by Kotani theory, the spectral type also changes completely from absolutely continuous spectrum to singular spectrum.

Note that the same paper also provides an average uniform convergence to the top Lyapunov exponents in the Oseledets theorem, which is an essential preparation for proving the regularity, namely Hölder continuity, of the Lyapunov exponent under generic assumptions.

4 Interlude: An Abstract LDT Theorem

As it is well known, one very powerful tool to proving the continuity of the Lyapunov exponent is the so-called large deviations type (LDT) theorem. In fact, Duarte and Klein [10] built an abstract continuity theorem (ACT) assuming uniform fiber LDT holds. Therefore, it is wise and natural to focus on LDT before going further.

In this section, we introduce an abstract LDT theorem for strongly mixing Markov systems obtained in [5]. Let us start by several basic definitions.

Markov Systems

Definition 4.1 A stochastic dynamical system (SDS) is any continuous map $K: M \to \mathrm{Prob}(M)$ on a Polish metric space M, where K is also called a Markov kernel.

Given any $x \in M$ and $E \subset M$ measurable, $K_x(E)$ represents the probability of going from x into E. The iterates are

$$K_x^{n+1}(E) = \int_M K_y^n(E) dK_x(y).$$

Definition 4.2 A probability measure $\nu \in \text{Prob}(M)$ is called a K-stationary if

$$\nu = K * \nu := \int K_x \, d\nu(x).$$

Definition 4.3 The Markov operator induced by an SDS $K\colon \mathcal{Q} = \mathcal{Q}_K : L^\infty(M, d\nu) \to L^\infty(M, d\nu)$ is defined by

$$(\mathcal{Q}\varphi)(x) := \int_M \varphi(y) \, dK_x(y).$$

Note that the powers of the Markov operator are given by

$$(Q^n \varphi)(x) = \int_M \varphi(y) dK_x^n(y)$$

Definition 4.4 A Markov system is a tuple $(M, K, \nu, \mathcal{E})$ where

(1) M is a Polish metric space,
(2) $K\colon M \to \text{Prob}(M)$ is an SDS,
(3) $\nu \in \text{Prob}(M)$ is a K-stationary measure,
(4) $\mathcal{E} = (\mathcal{E}, \|\cdot\|_{\mathcal{E}})$ is a Banach subspace of $L^\infty(M, d\nu)$ such that the inclusion $\mathcal{E} \subset L^\infty(M, d\nu)$ and the action of $\mathcal{Q}$ on $\mathcal{E}$ are both continuous. In other words there are constants $M_1 < \infty$ and $M_2 < \infty$ such that $\|\varphi\|_\infty \leq M_1 \, \|\varphi\|_{\mathcal{E}}$ and $\|\mathcal{Q}\varphi\|_{\mathcal{E}} \leq M_2 \, \|\varphi\|_{\mathcal{E}}$, for all $\varphi \in \mathcal{E}$.

Strong Mixing Now let us recall by order three definitions of strong mixing.
Firstly, the strongest version of mixing is:

Definition 4.5 (Meyn-Tweedie [14]) The Markov system is called uniformly ergodic if

$$\sup_{x \in M} \left\| K_x^n - \nu \right\|_{TV} \to 0 \quad \text{as} \quad n \to \infty$$

where $\|\cdot\|_{TV}$ refers to the total variation norm on the space of signed measures on M.

This condition is equivalent to the following: there exist some $C < \infty$ and $\sigma \in (0, 1)$ such that

$$\left\| \mathcal{Q}^n \varphi - \int_M \varphi \, d\nu \right\|_\infty \le C\sigma^n \, \|\varphi\|_\infty$$

for all $n \in \mathbb{N}$ and $\varphi \in L^\infty(M, d\nu)$.

Now let us state a weaker version of this concept.

Definition 4.6 (Duarte-Klein[10]) There are constants $C < \infty$ and $\sigma \in (0, 1)$ such that

$$\left\| \mathcal{Q}^n \varphi - \int_M \varphi \, d\nu \right\|_{\mathcal{E}} \le C \, \sigma^n \, \|\varphi\|_{\mathcal{E}}$$

for all $\varphi \in \mathcal{E}$ and $n \in \mathbb{N}$.

Note that the two definitions above lead to exponential LDT of the corresponding Markov system.

Finally, we may introduce the weakest version.

Definition 4.7 A Markov system $(M, K, \nu, \mathcal{E})$ is called strongly mixing with power mixing rate if there are constants $C < \infty$ and $p > 0$ such that for all $\varphi \in \mathcal{E}$ and $n \in \mathbb{N}$,

$$\left\| \mathcal{Q}^n \varphi - \int_M \varphi \, d\nu \right\|_\infty \le C \, \frac{1}{n^p} \, \|\varphi\|_{\mathcal{E}} \, .$$

Large Deviations Type Theorem As we know, the LDT for a measure-preserving dynamical system (MPDS) can be seen as a measure version of Birkhoff ergodic theorem. For all $\epsilon > 0$,

$$\nu \left\{ x \in M \colon \left| \frac{1}{n} \, S_n \, \varphi(x) - \int_M \varphi \, d\nu \right| > \epsilon \right\} \to 0 \quad \text{as } n \to \infty,$$

where $S_n\varphi := \varphi + \varphi \circ f + \cdots + \varphi \circ f^{n-1}$. Moreover, the precise convergence rate is quite important and usually only exponential and sub-exponential are useful for applications.

For a Markov system, we need a "stochastic" Birkhoff sum, and this is realized by Markov chain.

Let $X^+ = M^{\mathbb{N}}$, consider the sequence of random variables $\{Z_n : X^+ \to M\}_{n \in \mathbb{N}}$, $Z_n(x^+) := x_n$ where $x^+ = \{x_n\}_{n \in \mathbb{N}} \in X^+$.

By Kolmogorov, given $\pi \in \text{Prob}(M)$ there exists a unique probability measure $\mathbb{P}_\pi$ on X^+ for which $\{Z_n\}_{n \in \mathbb{N}}$ is a Markov chain with transition probability kernel K and initial probability distribution π, i.e., such that for every $x \in M$, every Borel set $A \subset M$ and any $n \ge 1$,

(a) $\mathbb{P}_\pi[\, Z_n \in A \mid Z_0, Z_1, \ldots, Z_{n-1} = x] = K_x(A)$,
(b) $\mathbb{P}_\pi[\, Z_0 \in A] = \pi(A)$.

When $\pi = \delta_x$, write $\mathbb{P}_x$.

When $\pi = \nu$, $\mathbb{P}_\nu$, called the Markov measure, is shift invariant and makes $\{Z_n\}_n$ a stationary process. Note that

$$\mathbb{P}_\nu(B) = \int_M \mathbb{P}_x(B)\, d\nu(x) \quad \text{and} \quad \mathbb{E}_\nu[\psi] = \int_M \mathbb{E}_x[\psi]\, d\nu(x)$$

for any Borel set $B \subset X^+$ and any bounded measurable function $\psi : X^+ \to \mathbb{R}$.

Let $\{Z_n\}_{n\geq 0}$ be the K-Markov chain: $Z_n \colon X^+ \to M$, $Z_n(x^+) = x_n$. For an observable $\varphi \colon M \to \mathbb{R}$ and an index $j \geq 0$ let

$$\varphi_j := \varphi \circ Z_j \colon X^+ \to \mathbb{R}.$$

Denote by

$$S_n\varphi := \varphi_0 + \cdots + \varphi_{n-1} = \varphi(Z_0) + \cdots + \varphi(Z_{n-1})$$

the corresponding "stochastic" Birkhoff sums.

Here comes the main theorem of this section.

Theorem 4.1 *Let $(M, K, \nu, \mathcal{E})$ be a strongly mixing Markov system with mixing rate $r_n = \frac{1}{n^p}$, $p > 0$. Then for all $\epsilon > 0$ and $\varphi \in \mathcal{E}$ there are $c(\epsilon) > 0$ and $n(\epsilon) \in \mathbb{N}$ such that for all $n \geq n(\epsilon)$ we have*

$$\mathbb{P}_\nu \left\{ \left| \frac{1}{n} S_n\varphi - \int_M \varphi d\nu \right| > \epsilon \right\} \leq 8e^{-c(\epsilon)n}$$

where $c(\epsilon) = c_1\, \epsilon^{2+\frac{1}{p}}$ and $n(\epsilon) = [c_2\, \epsilon^{-\frac{1}{p}}]$ for constants $c_1 > 0$ and $c_2 > 0$.

If the strongly mixing condition is with the uniform norm $\|\cdot\|_0$ on the left hand side, then the result holds with $\mathbb{P}_x$ for any $x \in M$.

The proof relies on the Bernstein's trick, the Hölder inequality and a trick of "sparse" rearrangement.

As an application, we proved an LDT (and also a CLT) for three levels of random torus translation realized by Markov chains for Hölder observables depending on the zero-th coordinate, the past coordinates and the full coordinates under a prevalent condition on the measure. For simplicity, we only present the LDT and CLT of the final level, which is exactly our base map (2.1).

Definition 4.8 We say that $\mu \in \text{Prob}(\mathbb{T}^d)$ satisfies a mixing Diophantine condition if there are positive constants γ and τ such that

$$|\hat{\mu}(k)| \leq 1 - \frac{\gamma}{\|k\|^\tau} \quad \forall k \in \mathbb{Z}^d \setminus \{0\}. \tag{4.1}$$

Let $\mathrm{DC}_m(\gamma, \tau)$ denote the set of all probability measures $\mu \in \mathrm{Prob}(\mathbb{T}^d)$ satisfying (4.1). Notice that $\mathrm{DC}_m(\gamma, \tau)$ is a compact and convex set of measures which is prevalent.

Theorem 4.2 *Assume that $\mu \in \mathrm{DC}_m(\gamma, \tau)$. Then for every observable $\varphi \in \mathcal{H}_\alpha(X \times \mathbb{T}^d)$ and $\epsilon > 0$, we have*

$$\mu^{\mathbb{Z}} \times m \left\{ \left| \frac{1}{n} S_n \varphi - \int \varphi d\mu^{\mathbb{Z}} \times m \right| > \epsilon \right\} < e^{-c(\epsilon)n}$$

where $c(\epsilon) = C\epsilon^{2+\frac{1}{p}}$ for some constant $C < \infty$ and $p = \frac{\alpha}{3\tau}-$.

Theorem 4.3 *Assume that $\mu \in \mathrm{DC}_m(\gamma, \tau)$ with $\alpha > 3\tau$. Then $\forall \varphi \in \mathcal{H}_\alpha(X \times \mathbb{T}^d)$ non-constant with zero mean, there exist $\sigma > 0$ depending on φ such that*

$$\frac{S_n \varphi}{\sigma \sqrt{n}} \xrightarrow{d} \mathcal{N}(0, 1) .$$

Here $\mathcal{H}_\alpha(X \times \mathbb{T}^d)$ denotes the space of α-Hölder continuous functions on $X \times \mathbb{T}^d$.

We note that Theorem 4.1, containing the best existing exponential LDT result, is not only interesting in itself, but also fundamental to our study because our mixed model can be embedded properly into a Markov system. Moreover, under certain assumptions we can prove that it is strongly mixing. This ensures the applicability of Theorem 4.1 and the regularity of LE is thus in our hands thanks to the ACT.

5 Regularity of the Lyapunov Exponent

In this section, we state two theorems for deterministic frequency and random-frequency mixed random-quasiperiodic cocycles which will appear in a forthcoming paper [6].

Let $X := \mathcal{G}^{\mathbb{Z}}$ be the space of sequences of quasiperiodic cocycles and denote by $\sigma : X \to X$ the usual two-sided shift.

Recall that a compactly supported measure $\nu \in \mathrm{Prob}_c(\mathcal{G})$ determines a *random-quasiperiodic cocycle* $F_\nu : X \times \mathbb{T}^d \times \mathbb{R}^m \to X \times \mathbb{T}^d \times \mathbb{R}^m$ by the formula

$$F_\nu(\omega, \theta, v) := (\sigma\omega, \theta + \alpha, A(\theta)\, v) ,$$

where $\omega_0 = (\alpha, A)$ is the 0-th coordinate of the sequence $\omega \in X$. The base of this cocycle is the map $f : X \times \mathbb{T}^d \to X \times \mathbb{T}^d$ defined by $f(\omega, \theta) := (\sigma\omega, \theta + \mathfrak{a}(\omega))$ where $\mathfrak{a} : X \to \mathbb{T}^d$, $\mathfrak{a}(\omega) := \alpha$, for $\omega_0 = (\alpha, A)$ as above. The map f preserves the measure $\nu^{\mathbb{Z}} \times m$, where m is the normalized Haar measure of $\mathbb{T}^d$. The matrix valued function of this cocycle is $\mathcal{A} : X \times \mathbb{T}^d \to \mathrm{SL}_m(\mathbb{R})$, defined by $\mathcal{A}(\omega, \theta) := A(\theta)$, where $\omega_0 = (\alpha, A)$.

We say that the random-quasiperiodic cocycle determined by $\nu \in \mathrm{Prob}_c(\mathcal{G})$ has a *deterministic frequency* $\alpha \in \mathbb{T}^d$ when $\mathrm{supp}(\mu) \subset \mathcal{G}_\alpha := \{\alpha\} \times C^1(\mathbb{T}^d, \mathrm{SL}_m(\mathbb{R}))$. In this case the measure $\nu \in \mathrm{Prob}_c(\mathcal{G}_\alpha)$ takes the form $\nu = \delta_\alpha \times \tilde{\nu}$ for some $\tilde{\nu} \in \mathrm{Prob}_c(C^1(\mathbb{T}^d, \mathrm{SL}_m(\mathbb{R})))$.

We say that $\alpha \in \mathbb{T}^d$ is *rationally independent* if $\langle k, \alpha \rangle \notin \mathbb{Z}$ for all $k \in \mathbb{Z}^d \setminus \{0\}$. Given $L < \infty$ denote by $C^1_L(\mathbb{T}^d, \mathrm{SL}_m(\mathbb{R}))$ the set of all functions $A \in C^1(\mathbb{T}^d, \mathrm{SL}_m(\mathbb{R}))$ such that $\|A\|_{C^1} \leq L$ and $\left\|A^{-1}\right\|_{C^1} \leq L$.

Finally let $L_1(\nu) \geq L_2(\nu) \geq \ldots \geq L_d(\nu)$ denote the Lyapunov exponents of the random-quasiperiodic cocycle determined by μ.

The main results of this section are

Theorem 5.1 *Given $0 < L < \infty$, $\alpha \in \mathbb{T}^d$ rationally independent and a measure $\tilde{\nu}_0 \in \mathrm{Prob}_c(C^1_L(\mathbb{T}^d, \mathrm{SL}_m(\mathbb{R})))$ assume that $\nu_0 := \delta_\alpha \times \tilde{\nu}_0$ satisfies:*

(1) ν_0 is quasi-irreducible;
(2) $L_1(\nu_0) > L_2(\nu_0)$.

Then locally near the meausre $\tilde{\nu}_0$ in $\mathrm{Prob}_c(C^1_L(\mathbb{T}^d, \mathrm{SL}_m(\mathbb{R})))$, the following hold

(a) The cocycles F_ν with $\nu := \delta_\alpha \times \tilde{\nu}$ and $\tilde{\nu} \in \mathcal{U}$, satisfy uniform LDT estimates of exponential type over $\mathcal{U}$.
(b) The function $\mathcal{U} \ni \tilde{\nu} \mapsto L_1(\nu)$, where $\nu := \delta_\alpha \times \tilde{\nu}$, is Hölder continuous w.r.t. the Wasserstein distance.

For any $L < \infty$ let $\Sigma_L := \mathbb{T}^d \times C^1_L(\mathbb{T}^d, \mathrm{SL}_m(\mathbb{R}))$.

Theorem 5.2 *Given positive constants L, γ and τ and a measure $\nu_0 \in \mathrm{Prob}_c(\Sigma_L)$ satisfying:*

(1) ν_0 is quasi-irreducible;
(2) $L_1(\nu_0) > L_2(\nu_0)$;
(3) $\mu_0 := \mathfrak{a}_* \nu_0 \in \mathrm{DC}_m(\gamma, \tau)$.

Then locally near the measure ν_0 in $\mathrm{Prob}_c(\Sigma_L)$, the following hold

(a) The cocycles F_ν with $\nu \in \mathcal{U}$, satisfy uniform LDT estimates of exponential type over $\mathcal{U}$.
(b) If τ is small enough, all cocycles F_ν with $\nu \in \mathcal{U}$, satisfy a CLT.
(c) The function $\mathcal{U} \ni \nu \mapsto L_1(\nu)$ is Hölder continuous w.r.t. the Wasserstein distance.

As explained previously, Theorems 5.1 and 5.2 will be consequences of the abstract LDT theorem. However, there is an essential difference between deterministic frequency and random frequencies. That is, in the deterministic frequency case, we do not have the "full" strong mixing of our Markov operator as the torus translation is never mixing. Luckily, we may separate the eigenspace into a direct sum of "quasi-periodic" subspace and a "random" one where the action of our Markov operator becomes a torus translation on the quasi-periodic subspace which has no loss of parameters (due to uniform convergence of Birkhoff sums by unique ergodicity) and thus LDT is immediate. For the random part, we can prove strong mixing so the LDT holds also. Overall, although the "full" strong mixing does not hold, the "full" LDT does hold.

The technical part after embedding our model into some suitable Markov system lies in proving the direct sum decomposition and the strong mixing of the Markov operator restricted to the random part (which does not mean that the random-frequency case is easier). In fact, lots of effort was devoted to both cases, involving many results from the previous papers [3–5].

Finally, Theorems 5.1 and 5.2 apply to the Schrödinger operators in Propositions 3.1 3.23.3, which again shows that randomness dominates quasi-periodicity in the sense of regularity of the Lyapunov exponent. For example, the Lyapunov exponent of a quasi-periodic cocycle can be discontinuous even with a very good topology [15]. However, as the randomness participates, the mixed cocycle exhibits Hölder continuous Lyapunov exponent consistent with the random case.

To finish this section, we note that Hölder continuity with respect to the measure is the strongest result as it implies the same result for the energy, the frequency and the potential about which many mathematicians in Mathematical Physics are more concerned.

6 Stability of the Lyapunov Exponent

With all the preparations in hand, we are finally ready to approach the original question of You.

Formulation of the Problem Given a rationally independent frequency α on the d-dimensional torus $\mathbb{T}^d = (\mathbb{R}/\mathbb{Z})^d$, let $\tau_\alpha(\theta) = \theta + \alpha \mod 1$ be the translation by α on $\mathbb{T}^d$ endowed with the Haar measure m.

Given a function $v\colon \mathbb{T}^d \to \mathbb{R}$, let $A_{0,v}\colon \mathbb{T}^d \to \mathrm{SL}_2(\mathbb{R})$,

$$A_{0,v}(\theta) := \begin{bmatrix} v(\theta) & -1 \\ 1 & 0 \end{bmatrix}$$

The skew-product map $F_{0,v}\colon \mathbb{T}^d \times \mathbb{R}^2 \to \mathbb{T}^d \times \mathbb{R}^2$,

$$F_{0,v}(\theta, p) := (\tau_\alpha(\theta), A_{0,v}(\theta)p)$$

is called a quasiperiodic Schrödinger cocycle. We will also refer to the matrix valued function $A_{0,v}$ as being the same quasiperiodic cocycle.

Let $\rho \in \mathrm{Prob}_c(\mathbb{R})$ and consider the following random perturbation of the quasiperiodic cocycle A_0:

$$A_{\epsilon,v}(\omega_0, \theta) := \begin{bmatrix} v(\theta) + \epsilon\,\omega_0 & -1 \\ 1 & 0 \end{bmatrix} = P(\epsilon\,\omega_0)\, A_{0,v}(\theta),$$

where

$$P(\omega_0) := \begin{bmatrix} 1 & \omega_0 \\ 0 & 1 \end{bmatrix}$$

Denote by Σ the support of ρ, let $X := \Sigma^{\mathbb{Z}}$ and consider the two-sided Bernoulli shift σ on $(X, \rho^{\mathbb{Z}})$. The random perturbation $A_{\epsilon,v}$ determines the linear cocycle $F_{\epsilon,v} \colon X \times \mathbb{T}^d \times \mathbb{R}^2 \to X \times \mathbb{T}^d \times \mathbb{R}^2$,

$$F_{\epsilon,v}(\omega, \theta, p) := (\sigma\omega, \theta + \alpha, A_{\epsilon,v}(\omega, \theta)p)$$

over the mixed random-quasiperiodic base dynamics

$$X \times \mathbb{T}^d \ni (\omega, \theta) \mapsto f(\omega, \theta) = (\sigma\omega, \tau_\alpha(\theta)) .$$

For this Schrödinger case, we have the following theorem.

Theorem 6.1 *Assume the function $v \colon \mathbb{T}^d \to \mathbb{R}$ is analytic, the frequency $\alpha \in \mathbb{T}^d$ satisfies a Diophantine condition, $L_1(A_{0,v}) > 0$ and $\rho \in \mathrm{Prob}_c(\mathbb{R})$ has positive pointwise Hausdorff dimension, i.e., for some constants $C < \infty$ and $\vartheta > 0$*

$$\rho(B(x, r)) \le C\, r^{\vartheta} \quad \forall x \in \mathbb{R}, r > 0.$$

Then the map

$$(\epsilon, v) \mapsto L_1(F_{\epsilon,v})$$

is locally weak-Hölder continuous on $[0, \infty) \times C^{\omega}_r(\mathbb{T}^d, \mathbb{R})$. In particular the Lyapunov exponent is stable under random perturbations.

This theorem is a particular version of the following general result.

Definitions and General Statement Let $\mathrm{sl}_m(\mathbb{R})$ denote the Lie algebra of the group $\mathrm{SL}_m(\mathbb{R})$ and consider a setting with the following four ingredients:

(1) $\alpha \in \mathbb{T}^d$ a frequency satisfying a Diophantine condition;
(2) $A_0 \in C^{\omega}_r(\mathbb{T}^d, \mathrm{SL}_m(\mathbb{R}))$ an analytic cocycle;
(3) $\mu \in \mathrm{Prob}_c(\mathrm{sl}_m(\mathbb{R}))$ a compactly supported measure.

These determine the following objects: A Bernoulli space of sequences $\left(X := \mathrm{sl}_m(\mathbb{R})^{\mathbb{Z}}, \mu^{\mathbb{Z}}\right)$ endowed with a two-sided Bernoulli shift $\sigma : X \to X$. With it we define the base map $f : X \times \mathbb{T}^d \to X \times \mathbb{T}^d$, $f(\omega, \theta) := (\sigma\omega, \theta + \alpha)$, which together with the matrix valued function $A_\epsilon : X \times \mathbb{T}^d \to \mathrm{SL}_m(\mathbb{R})$, $A_\epsilon(\omega, \theta) := e^{\epsilon\,\omega_0} A_0(\theta)$ determines the randomly perturbed linear cocycle $F_{\alpha,A_0,\mu,\epsilon} : X \times \mathbb{T}^d \times \mathbb{R}^m \to X \times \mathbb{T}^d \times \mathbb{R}^m$,

$$F_{\alpha,A_0,\mu,\epsilon}(\omega, \theta, p) := (f(\omega, \theta), A_\epsilon(\omega, \theta)\, p).$$

We make two assumptions. First we assume a gap on the Lyapunov spectrum.

(H1) $L_1(\alpha, A_0) > L_2(\alpha, A_0)$.

The second hypothesis is a bit more technical.

(H2) There exist constants $C < \infty$, $\vartheta > m - 2$, $\varepsilon_0 > 0$ and $k_0 \in \mathbb{N}$ such that for every $\theta \in \mathbb{T}^d$, $\hat{p}, \hat{q} \in \mathbb{P}(\mathbb{R}^m)$ and $0 < r < \varepsilon < \varepsilon_0$

$$\mu^{\mathbb{Z}} \left\{ \omega \in X\colon \hat{A}_{\varepsilon_0}^{k_0}(\omega, \theta)\, \hat{p} \in B(\hat{q}, r) \right\} < C \left(\frac{r}{\varepsilon} \right)^{\vartheta}.$$

Then we have a more general theorem as follows.

Theorem 6.2 *Fix* $\alpha \in \mathbb{T}^d$ *satisfying a Diophantine condition and the space* $\mathcal{U} \subseteq \mathrm{Prob}(\mathrm{sl}_m(\mathbb{R})) \times C_r^{\omega}(\mathbb{T}^d, \mathrm{SL}_m(\mathbb{R})) \times [0, +\infty)$ *of all tuples* (μ, A_0, ϵ) *such that* (μ, A_0) *satisfy hypothesis (H1) and (H2). Then the function* $L_1 : \mathcal{U} \to \mathbb{R}$, $(\mu, A_0, \epsilon) \mapsto L_1(F_{\alpha, A_0, \mu, \epsilon})$ *is locally weak-Hölder continuous.*

The proof is based on an approximation lemma by quasi-periodic estimates, results of mixed random-quasiperiodic cocycles and a bridging argument by the Avalanche Principle (AP). The key idea is that for a small number of iterates, the system is behaving more like the quasi-periodic one while if there are enough iterates, the randomness starts to work and dominate. Finally, by AP the gap between these small and large numbers of iterates can be perfectly filled in!

We believe that (H2) could be removed though at this point we cannot. That is why we call our result partial. In fact, there are many other questions that we are trying to resolve. For example, stability in the random-frequency case and metal-insulator transition from quasi-periodic models to mixed models, etc.

Hopefully, the papers in preparation will be released soon.

Last but not least:

Acknowledgments The author would like to give his deepest and sincerest gratitude to Pedro Duarte, Silvius Klein, Jiangong You and Qi Zhou for their consistent support and persistent inspiration.

References

1. Ao Cai, *The absolutely continuous spectrum of finitely differentiable quasi-periodic Schrödinger operators*, Ann. Henri Poincaré **23** (2022), 4195–4226.
2. Ao Cai, Claire Chavaudret, Jiangong You, and Qi Zhou, *Sharp Hölder continuity of the Lyapunov exponent of finitely differentiable quasi-periodic cocycles*, Math. Z. **291** (2019), no. 3-4, 931–958. MR 3936094
3. Ao Cai, Pedro Duarte, and Silvius Klein, *Furstenberg theory of mixed random-quasiperiodic cocycles*, Commun. Math. Phys. **402** (2023), 447–487.
4. Ao Cai, Pedro Duarte, and Silvius Klein, *Mixed Random-Quasiperiodic Cocycles*, Bull. Braz. Math. Soc. (N.S.) **53** (2022), no. 4, 1469–1497. MR 4502841

5. Ao Cai, Pedro Duarte, and Silvius Klein, *Statistical properties for mixing Markov chains with applications to dynamical systems*, 2022.
6. Ao Cai, Pedro Duarte, and Silvius Klein, *Hölder continuity of the Lyapunov exponent for mixed random-quasiperiodic cocycles*, In preparation.
7. Ao Cai, Pedro Duarte, and Silvius Klein, *Stability of Lyapunov exponents of quasiperiodic linear cocycles under random noise*, In preparation.
8. Ao Cai and Lingrui Ge, *Reducibility of finitely differentiable quasi-periodic cocycles and its spectral applications*, J. Dyn. Diff. Equat. **34** (2022), 2079–2104.
9. Ao Cai and Xueyin Wang, *Polynomial decay of the gap length for C^k quasi-periodic Schrödinger operators and spectral application*, J. Funct. Anal. **281** (2021), no. 3, 109035.
10. Pedro Duarte and Silvius Klein, *Lyapunov exponents of linear cocycles*, Atlantis Studies in Dynamical Systems, Atlantis Press, 2016.
11. Pedro Duarte and Silvius Klein, *Continuity of the Lyapunov exponents of linear cocycles*, Publicações Matemáticas, 31° Colóquio Brasileiro de Matemática, IMPA, 2017, available at https://impa.br/wp-content/uploads/2017/08/31CBM_02.pdf.
12. Alex Furman, *Random walks on groups and random transformations*, Handbook of dynamical systems, Vol. 1A, North-Holland, Amsterdam, 2002, pp. 931–1014.
13. Harry Furstenberg, *Noncommuting random products*, Trans. Amer. Math. Soc. **108** (1963), 377–428.
14. Sean Meyn and Richard L. Tweedie, *Markov chains and stochastic stability*, second ed., Cambridge University Press, Cambridge, 2009, With a prologue by Peter W. Glynn. MR 2509253
15. Yiqian Wang and Jiangong You, *Examples of discontinuity of Lyapunov exponent in smooth quasiperiodic cocycles*, Duke Math. J. **162** (2013), no. 13, 2363–2412.

Hyperbolicity or Zero Lyapunov Exponents for C^2-Hamiltonians

João Lopes Dias and Filipe Santos

Abstract Hamiltonian systems on compact symplectic manifolds form a rich class of dynamical systems with many applications. We describe a proof that for a C^2-residual set of Hamiltonians there are distinct dynamics in the spirit of the Bochi-Mañé dichotomy: uniform hyperbolicity or zero Lyapunov exponents.

1 Introduction

There is a natural interest in understanding the dependence of Lyapunov exponents with respect to variations of the underlying dynamical system. For C^1 conservative diffeomorphisms on compact surfaces, Mañé [8, 9] conjectured and gave key ideas to show that, generically, non-Anosov maps have zero Lyapunov exponents almost everywhere. This was later proved by Bochi in [4], and it is now known as the Bochi-Mañé dichotomy. Later, Bochi and Viana [7] proved versions of this dichotomy for volume preserving diffeomorphisms in any dimension and Bochi extended the results to the symplectic case [5]. The goal of this work is to present the Hamiltonian version contained in [10]. Notice that a Bochi-Mañé dichotomy for Hamiltonian flows on 4-dimensional compact symplectic manifolds already appeared in [2] and for 3-dimensional volume-preserving flows in [1].

Consider a smooth compact symplectic $2d$-dimensional manifold (M, ω). The set $C^2(M)$ of Hamiltonians consists of C^2 real functions on M, constant on each connected component of ∂M. We restrict to the C^2-topology since our perturbation techniques do not work on more refined topologies. A reason for this is that we need to rescale the perturbation in order to assure that its support is contained in a small enough neighborhood of a segment of the orbit and does not interfere with

J. L. Dias (✉)
Departamento de Matemática, CEMAPRE/REM/ISEG, Universidade de Lisboa, Lisboa, Portugal
e-mail: jldias@iseg.ulisboa.pt

F. Santos
Departamento de Matemática, ISEG, Universidade de Lisboa, Lisboa, Portugal

© The Author(s), under exclusive license to Springer Nature Switzerland AG 2023
J. L. Dias et al. (eds.), *New Trends in Lyapunov Exponents*, CIM Series in Mathematical Sciences, https://doi.org/10.1007/978-3-031-41316-2_4

the perturbations elsewhere. This rescaling, despite maintaining the derivative of the flow, leads to the uncontrolled growth of the second derivative of the flow, affecting the C^3 estimates of the Hamiltonians.

Let $H \in C^2(M)$, the corresponding Hamiltonian vector field X_H and Hamiltonian flow φ_H^t. The volume form ω^n gives a measure μ in M preserved by φ_H^t.

Theorem 1.1 *For a C^2-residual subset of Hamiltonians in $C^2(M)$ there is an invariant decomposition $M = D \cup Z$ μ-mod 0 such that:*

- *D is a countable increasing union of compact invariant sets with a dominated splitting, and*
- *for each $x \in Z$ the Hamiltonian flow has all Lyapunov exponents equal to zero.*

This theorem implies [2, Theorem 2] for $d = 2$. Recall that any invariant set with dominated splitting is partially hyperbolic [6, Theorem 11]. So, Theorem 1.1 can be stated as a dichotomy between partial hyperbolicity and zero Lyapunov exponents.

2 Basic Definitions

For each energy level set

$$H^{-1}(e) = \{x \in M \colon H(x) = e\},$$

we denote a connected component by $\mathcal{E} \subset H^{-1}(e)$. The Hamiltonian vector field X_H is tangent to each $H^{-1}(e)$, and so these sets are invariant under the Hamiltonian flow. The residual character of Morse functions implies the existence of only finitely many critical points and we will therefore focus only on connected components of regular energy levels, called energy (hyper) surfaces $\mathcal{E}$. One can also obtain a measure $\mu_{\mathcal{E}}$ corresponding to the restriction of μ to $\mathcal{E}$. This is related to the volume form $\omega^n(X, \cdot)$ on the tangent space $T_x\mathcal{E}$, where X is a vector field transversal to $\mathcal{E}$.

Let $H \in C^2(M)$ and $\mathcal{O} \subset M$ be the full μ-measure set of Oseledets regular points for φ_H^t. Assuming that the Oseledets splitting at $x \in \mathcal{O}$ is not trivial, we have

$$T_x M = E_x^1 \oplus E_x^2 \oplus \cdots E_x^{k(x)} \oplus \mathbb{R} X_H(x),$$

with $1 \leq k(x) \leq 2d - 1$, corresponding to the Lyapunov exponents

$$\lambda_1(H, x) \geq \cdots \geq \lambda_{2d}(H, x)$$

repeated with multiplicity $\dim E_x^j$. This splitting can be zipped by grouping together the subspaces with positive Lyapunov exponent on E^+, with zero exponent on E^0

and with negative exponent on E^-, into

$$T_x M = E_x^- \oplus E_x^0 \oplus E_x^+.$$

Note that the subspace E_x^0 includes $\mathbb{R}X_H(x)$, the Hamiltonian vector field direction. Moreover, $\lambda_i(H, x) = -\lambda_{d+i}(H, x)$, $1 \leq i \leq d$, as this is a symplectic system.

Following [2], associated to a Hamiltonian H we denote the transversal linear Poincaré flow by

$$\Phi_H^t(x)\colon \mathcal{N}_x \to \mathcal{N}_{\varphi_H^t(x)},$$

which acts on projected normal spaces $\mathcal{N}_x$ restricted to energy surfaces (dimension $2d - 2$). For each t, this is a linear symplectomorphism for the symplectic form induced on $\mathcal{N}_x$ by ω. The Lyapunov exponents are the same for Φ_H^t and φ_H^t except for the time and energy directions (having zero exponents). The zipped Oseledets splitting is then written in the suggestive notation:

$$\mathcal{N}_x = \mathcal{N}_x^- \oplus \mathcal{N}_x^0 \oplus \mathcal{N}_x^+.$$

Let $\Lambda \subset M$ be φ_H^t-invariant and $m > 0$. A splitting of the bundle $\mathcal{N}_\Lambda = E_\Lambda \oplus F_\Lambda$ is m-dominated if it is Φ_H^t-invariant, the subbundles E_Λ and F_Λ have constant dimension and

$$\frac{\|\Phi_H^m(x)|E_x\|}{\|\Phi_H^m(x)|F_x\|} \leq \frac{1}{2}, \qquad x \in \Lambda.$$

It is dominated if it is m-dominated for some m (the index of the splitting is the dimension of F_x). In this case, we write $E \to F$, which has the interpretation in the projective space that E is reppeling and F attracting. When there is no such domination, we write $E \not\to F$.

Notice that an invariant and compact set Λ is partially hyperbolic iff $\dim(\mathcal{N}^0) \geq 2$ and $\mathcal{N}^- \oplus \mathcal{N}^0 \oplus \mathcal{N}^+$ is a dominated zipped Oseledets splitting [7]. That is, $\mathcal{N}^{-0} \oplus \mathcal{N}^+$ and $\mathcal{N}^{-0} := \mathcal{N}^- \oplus \mathcal{N}^0$ are both dominated. So, $\mathcal{N}^- \to \mathcal{N}^0 \to \mathcal{N}^+$. Uniform hyperbolicity corresponds to $\mathcal{N}^0 = \{0\}$ and $\mathcal{N}^- \to \mathcal{N}^+$. Non-dominance means that $\mathcal{N}^{-0} \not\to \mathcal{N}^+$ and $\mathcal{N}^- \not\to \mathcal{N}^{0+}$.

It is known that if a set Λ inside an energy surface $\mathcal{E}$ is uniformly hyperbolic, then $\mu_{\mathcal{E}}(\Lambda) = 0$ or $\Lambda = \mathcal{E}$ (i.e. $\mathcal{E}$ is Anosov); see references in [2].

Consider a finite segment γ of the orbit $\varphi_H^t(x)$ of $x \in M$ for some H, corresponding to an interval $t \in [\tau, \tau + T]$. The corresponding curve of transversal linear Poincaré maps $\{\Phi_H^t(x)\}_{\tau < t < \tau + T}$ is flexible with respect to some given invariant subspaces E and F of the zipped Oseledets splitting, if we can find a C^2-perturbed Hamiltonian $\widetilde{H}$, compactly supported on a flow tube around γ, such that $\Phi_{\widetilde{H}}^{\tau+T}(x)E$ and F are arbitrarily close. Moreover, this must work for most points (in a μ-measure sense) in the support of the perturbation.

3 Sketch of the Proof

Suppose $x \in \mathcal{O}$ has at least one positive Lyapunov exponent for $\Phi_H^t(x)$ and the zipped Oseledets splitting is not dominated. The following steps relating dominance with flexibility are used to prove Theorem 1.1 (see [2, 10] for details).

1. Absence of a dominated zipped Oseledets splitting can only be of four distinct types (see [5]).
2. Non-dominance implies flexibility.
3. If an orbit supports a flexible sequence, we can perturb the Hamiltonian to create a local discontinuity of the Lyapunov exponent.

We start by approximating a Hamiltonian by a nearby smooth one $H \in C^\infty(M)$ in the C^2-topology, since $C^\infty(M)$ is dense. This is done to apply the Hamiltonian flowbox theorem around a segment of an orbit, with the coordinate transformation being also C^∞ (otherwise, the regularity would drop).

Now, the first three abovementioned types of non-dominance are dealt in a similar way as in [2], corresponding to $\mathcal{N}^- \not\rightarrow \mathcal{N}^+$ for any $\mathcal{N}^0$. The idea is to mix the directions $\mathcal{N}^-$ and $\mathcal{N}^+$ by rotations along planes of conjugated coordinates, in order to obtain flexible transversal linear Poincaré flows.

The fourth type is more complex as explained in [5]. It corresponds to the situation when $\mathcal{N}^{-0} \not\rightarrow \mathcal{N}^+$ and $\mathcal{N}^- \not\rightarrow \mathcal{N}^{0+}$ but $\mathcal{N}^- \rightarrow \mathcal{N}^+$. The local case can be dimensionally reduced to $\mathbb{R}^4$, with $\dim \mathcal{N}_x^- = \dim \mathcal{N}_x^+ = 1$ and $\dim \mathcal{N}_x^0 = 2$. In order to show flexibility, we perturb the Hamiltonian hoping that $\mathcal{N}^0$ and $\mathcal{N}^-$ (or $\mathcal{N}^+$) can be mixed for most of the points, even though $\mathcal{N}^- \rightarrow \mathcal{N}^+$. The challenge is to make it symplectically, as the planes do not correspond to canonical conjugated directions. It is enough to use a Hamiltonian suspension of a perturbation of the Poincaré map (an extension to [3] which is for periodic orbits) taken to be Bochi's symplectomorphism obtained via his original random perturbative idea [5].

Write the integrated Lyapunov exponent with index $p \in \{1, \ldots, d\}$ for a H and an invariant set $\Lambda \subset M$ as

$$LE_p(H, \Lambda) = \int_\Lambda \sum_{i=1}^{p} \lambda_i(H, x)\, d\mu(x),$$

where $\lambda_i(H, x)$ are the largest p positive Lyapunov exponents of H at x (repeated with multiplicity of the dimension of the corresponding Oseledets subspace). Define also

$$J_p(H, \Lambda) = \int_{\Gamma_p \cap \Lambda} \frac{\lambda_p(H, x) - \lambda_{p+1}(H, x)}{2}\, d\mu(x),$$

where $\Gamma_p := \bigcap_m \Gamma_{m,p}(H)$ and $\Gamma_{m,p}(H)$ denotes the set of points in M without m-dominated splitting of index p.

Flexibility implies a jump of the Lyapunov exponent because it allows to mix invariant directions of the zipped Osedelets splitting, for a positive measure set of points. After globalization, using a Kakutani tower scheme, one obtains the following result.

Proposition 3.1 ([10]) *Let* $H \in C^{\infty}(M)$ *and* $\epsilon, \delta > 0$*. Then there exists* $\widetilde{H} \in C^2(M)$*,* ϵ*-*C^2*-close to* H *and equal to* H *on* $M \setminus \cup_p \Gamma_p$*, such that*

$$LE_p(\widetilde{H}, \Lambda) < LE_p(H, \Lambda) + \delta - J_p(H, \Lambda).$$

It is known that the map $H \mapsto LE_p(H, \Lambda)$ is upper semi-continuous for Hamiltonians having a common invariant set Λ (follows from similar ideas to the ones in [7]). Thus, the set of its continuity points is a residual set. Notice that discontinuities correspond to $J_p(H, \Lambda) > 0$ for a good choice of Λ.

As in the statement of Theorem 1.1, let $D \subset M$ be the set of points with a dominated splitting for H and Z the points with all exponents equal to zero. We want to show that if $H \in C^2(M)$ is a point of continuity of $LE_p(\cdot, \Lambda)$ for any $p \in \{1, \dots, d\}$, then $\mu(D \cup Z) = 1$.

By a small perturbation we assume that $H \in C^{\infty}(M)$. Suppose that $\mu(D \cup Z) < 1$, i.e. $\mu(\Gamma_p \cap Z^c) > 0$ for some p. This means that $J_p(H, Z^c) > 0$ since p is such that $\lambda_p(H, x) > \lambda_{p+1}(H, x)$. By Proposition 3.1, H is not a continuity point.

In conclusion, for a residual set of Hamiltonians, a.e. x either has a dominated splitting or all the exponents are equal to zero.

Acknowledgments The authors were partially funded by the projects 'New trends in Lyapunov exponents' PTDC/MAT-PUR/29126/2017 and CEMAPRE—UID/MULTI/00491/2019, both financed by Fundação para a Ciência e a Tecnologia, Portugal.

References

1. M. Bessa. The Lyapunov exponents of generic zero divergence 3-dimensional vector fields. *Ergod. Th. Dynam. Sys.*, 27:1445–1472, 2007.
2. M. Bessa and J. Lopes Dias. Generic dynamics of 4-dimensional C^2 Hamiltonian systems. *Commun. Math. Phys.*, 281:597–619, 2008.
3. Mário Bessa and João Lopes Dias. Hamiltonian suspension of perturbed Poincaré sections and an application. *Math. Proc. Cambridge Philos. Soc.*, 157(1):101–112, 2014.
4. J. Bochi. Genericity of zero Lyapunov exponents. *Ergod. Th. Dynam. Sys.*, 22:1667–1696, 2002.
5. J. Bochi. C^1-generic symplectic diffeomorphisms: partial hyperbolicity and zero center Lyapunov exponents. *Journal of the Inst. of Math. Jussieu*, 9:49–93, 2010.
6. J. Bochi and M. Viana. *Advances in Dynamical Systems*, chapter Lyapunov exponents: How frequently are dynamical systems hyperbolic? Cambridge Univ. Press, 2004.
7. J. Bochi and M. Viana. The Lyapunov exponents of generic volume preserving and symplectic maps. *Ann. Math.*, 161:1423–1485, 2005.

8. Ricardo Mañé. Oseledec's theorem from the generic viewpoint. In *Proceedings of the International Congress of Mathematics, Warszawa*, volume 2, pages 1259–1276, 1983.
9. Ricardo Mañé. The Lyapunov exponents of generic area preserving diffeomorphisms. In *Conference on Dynamical Systems, Montevideo*, 1995.
10. Filipe Santos. *Bochi-Mañé Dichotomy for* $2n$*-Hamiltonians, Random Perturbation Techniques*. PhD thesis, ISEG - Universidade de Lisboa, 2020.

Generalized Lyapunov Exponents and Aspects of the Theory of Deep Learning

Anders Karlsson

Abstract We discuss certain recent metric space methods and some of the possibilities these methods provide, with special focus on various generalizations of Lyapunov exponents originally appearing in the theory of dynamical systems and differential equations. These generalizations appear for example in topology, group theory, probability theory, operator theory and deep learning.

1 Introduction

The law of large numbers states that for a sequence of independent, identically distributed (i.i.d.) random variables X_1, X_2, ..., X_n with finite expectation,

$$(X_1 + X_2 + ... + X_n)/n \to \mathbb{E}[X_1] \tag{1}$$

almost surely as $n \to \infty$. Bellman [5], Furstenberg [21] and others asked whether in some situations there could exist a similar limit law for products

$$u(n) := g_1 g_2 g_3 ... g_n$$

of i.i.d. noncommutative operations g_1, g_2, ..., g_n. Such products appear for example as solutions to difference equations with random coefficients, or from time-one maps of the solutions of continuous models, say from a stochastic PDE. In addition to mathematics and physics (early references being [15, 70]), one can find papers in biology, epidemiology, medicine, and economics leading to random products of noncommuting transformations [7, 12, 33, 62]. Compositional products is also one of the key features of deep learning as will be highlighted below.

A. Karlsson (✉)
Section de mathématiques, Université de Genève, Genève, Switzerland

Mathematics Department, Uppsala University, Uppsala, Sweden
e-mail: anders.karlsson@unige.ch

© The Author(s), under exclusive license to Springer Nature Switzerland AG 2023
J. L. Dias et al. (eds.), *New Trends in Lyapunov Exponents*, CIM Series in Mathematical Sciences, https://doi.org/10.1007/978-3-031-41316-2_5

Note that in contrast to (1) it is unclear how to form an average in the noncommutative setting. Important partial answers to the above question were obtained at the end of the 1960s ([48, 63]) and later as one aspect of random walks on groups, see for example [6, 18, 26, 36, 46, 55, 77]. A quite general affirmative answer to the question of a limit law for noncommuting random products was provided in [25, 43], see Theorem 3 below.

In ergodic theory one formalizes the setting as follows, more general than the i.i.d. assumption. Let (Ω, μ) be a measure space with $\mu(\Omega) = 1$. Let $T : \Omega \to \Omega$ be a measurable map preserving the measure. We furthermore assume ergodicity, which is an irreducibility assumption that states that up to measure zero there are no T-invariant subsets of Ω. Given a measurable map $g : \Omega \to G$ (assigning some measurable structure on the group G; g is what a probabilist would call a random variable), we define the following *ergodic cocycle*:

$$u(n, \omega) := g(\omega)g(T\omega)...g(T^{n-1}\omega).$$

In addition, one needs to assume that the cocycle is integrable which means that the integral over Ω of the "size" of g is finite.

For matrices, a first answer was provided by Furstenberg-Kesten for the norm of the matrices, and a more precise answer was given later in the 1960s by Oseledets in his multiplicative ergodic theorem. One can view this as a random spectral theorem, intuitively it says that the random product behaves in the same way as the powers of one single "average" matrix:

Theorem 1 (Oseledets' Multiplicative Ergodic Theorem [63]) *Let* $A(n, \omega) = g_n g_{n-1}...g_1$ *be an integrable ergodic cocycle of invertible matrices. Then there are a.s. a random filtration of subspaces* $0 = V_0 \subset V_1 \subset ... \subset V_k = \mathbb{R}^d$ *and numbers* $\lambda_1 < \lambda_2 < ... < \lambda_k$ *such that*

$$\lim_{n\to\infty} \frac{1}{n} \log \|A(n, \omega)v\| = \lambda_i$$

whenever $v \in V_i \setminus V_{i-1}$.

The case $d = 1$ is the Birkhoff ergodic theorem generalizing (1). The numbers λ_i are called *Lyapunov exponents.* This is a fundamental theorem in the theory of differentiable dynamical systems and has also many other applications, the most spectacular such is Margulis' proof [56] of his super-rigidity theorem. As for physics, Nobel laureate Parisi wrote in the foreword of [13] that "The properties of random matrices and their products form a basic tool, whose importance cannot be underestimated. They play a role as important as Fourier transforms for differential equations."

Now compare the above with the following. Let Σ be an oriented closed surface of genus $g \geq 2$. Let S denote the isotopy classes of simple closed curves on M not isotopically trivial. For a Riemannian metric ρ on Σ, let $l_\rho(\beta)$ be the infimum of the length of curves isotopic to β. In a legendary preprint from 1976 [75], Thurston announced the following (the details are worked out in [20, Théorème Spectrale] using foliation theory):

Theorem 2 (Thurston's Spectral Theorem for Surface Diffeomorphisms [75]) *Let f be a diffeomorphism of a surface Σ of genus $g \geq 2$. Then there is filtration of subsurfaces $Y_1 \subset Y_2 \subset \ldots \subset Y_k = \Sigma$ and algebraic integers $\lambda_1 < \lambda_2 < .. < \lambda_k$ such that*

$$\lim_{n\to\infty} \frac{1}{n} \log l_\rho(f^n c) = \lambda_i$$

whenever the simple closed curve c can be isotoped to a curve contained in Y_i but not in Y_{i-1}.

This is analogous to a simple statement for linear transformations A in finite dimensions (which corresponds to the Oseledets theorem in the case $A(n, \omega) = A^n$): given a vector v there is an associated exponent λ (absolute value of an eigenvalue), such that

$$\lim_{n\to\infty} \left\| A^n v \right\|^{1/n} = \lambda.$$

To spell out the analogy: a diffeomorphism f instead of a linear transformation A, a length instead of a norm, and a curve α instead of a vector v. And the answer is given in similar terms: Lyapunov exponents and associated filtration of subspaces and subsurfaces respectively.

A *weak metric space* (following the terminology of for example [27]) is a set X equipped with a function $X \times X \to \mathbb{R}$ such that

$$d(x, x) = 0$$

and

$$d(x, y) < d(x, z) + d(z, y)$$

for all points $x, y, z \in X$. A map $f : X \to X$ is called *nonexpansive* if

$$d(f(x), f(y)) \leq d(x, y)$$

for all $x, y \in X$.

The following multiplicative ergodic theorem was proved for isometries in [43] and in general in [25].

Theorem 3 (Ergodic Theorem for Noncommuting Random Products, [25, 43]) *Let $u(n, \omega)$ be an integrable ergodic cocycle of nonexpansive maps of a weak metric space (X, d) and such that $\omega \mapsto u(n, \omega)x$ is measurable. Then there exists a.s. a metric functional h of X such that*

$$\lim_{n\to\infty} -\frac{1}{n} h(u(n, \omega)x) = \lim_{n\to\infty} \frac{1}{n} d(x, u(n, \omega)x).$$

Fig. 1 Illustration of the convergence in direction of $u(n, \omega)x$ as $n \to \infty$ in six experiments in a hyperbolic disk. The colored disks are horodisks that the random walks trajectories go deeper and deeper into with time (Image by Cécile Bucher)

Metric functionals are almost what is usually called horofunctions, see Sect. 3 and Fig. 1. This theorem when specialized to X a symmetric space of nonpositive curvature, Gromov hyperbolic space, or CAT(0) space, recovers some previous results mentioned above by Oseledets, Furstenberg, Kaimanovich, and Karlsson-Margulis. It also implies random mean ergodic theorems of Ulam-von Neumann, Kakutani, and Beck-Schwartz, see [25], and it holds even when the traditional mean ergodic theorem fails [41]. Theorem 3 furthermore provides generalized laws of large numbers with concave moments [47], and has found application to random walks on groups and bounded harmonic functions on manifolds [44, 45] without knowing anything specific about the metric functionals in these settings. I want to emphasize that Theorem 3 applies in particular to every random walks with finite first moment on *any* finitely generated group.

Moreover, using Theorem 3, Horbez, building on the approach of [39], could establish the following random extension of Thurston's theorem:

Theorem 4 (Random Spectral Theorem of Surface Homeomorphisms [32, 39]) *Let $v(n, \omega) = A(T^{n-1}\omega)...A(T\omega)A(\omega)$ be an integrable i.i.d random product of homeomorphisms of a closed surface Σ of genus $g \geq 2$. Then there is a (random) filtration of subsurfaces $Y_1 \subset Y_2 \subset ... \subset Y_k = \Sigma$ and (deterministic) exponents $\lambda_1 < \lambda_2 < .. < \lambda_k$ such that*

$$\lim_{n\to\infty} \frac{1}{n} \log l_\rho(v(n, \omega)c) = \lambda_i$$

whenever the simple closed curve c can be isotoped to a curve contained in Y_i *but not in* Y_{i-1}*. Here* l_ρ *is the minimal length in the isotopy class in some fixed Riemannian metric* ρ.

The exponents λ_i are a type of generalized Lyapunov exponents that perhaps could be called *topological Lyapunov exponents for surface homeomorphisms.* A different approach was provided in [40] which showed how to get the top exponent for ergodic cocycles of homeomorphisms, using the metric ideas and a lemma in [46] combining it with results in [51]. In both cases, the proofs use Thurston's asymmetric metric. A prior study of random walks on the mapping class groups was carried out in [35], which showed in particular that under a non-elementary assumption the random walk converges to uniquely ergodic foliations, in which case it follows from [39] that there is only one exponent.

Actually the main results of Horbez' paper [32] concern instead random walks on the outer automorphism group of free groups, giving a result very similar to Theorem 4. In order not to have to explain notations from the important subject of automorphisms of free groups, I will not state it here. The proof goes via a determination of the metric functionals of the outer space and an application of Theorem 3. To get all generalized Lyapunov exponents Horbez then studies the set of stationary measures on the boundary of outer space in parallel to works in the matrix case of Furstenberg-Kifer and Hennion.

We thus see three settings, linear transformations in finite dimensions, surface homeomorphisms, and automorphisms of free groups, that are not merely analogous but the corresponding "law of large numbers" can be deduced ultimately from the same theorem. The strategy is:

- Instead of looking at the underlying space where the linear maps, homeomorphisms, group automorphisms etc act, we lift the action to a more abstract space, a *moduli space* as it were, of positive structures on the corresponding underlying space.
- On that associated space there is often an invariant metric.
- Employ the noncommutative ergodic theorem in terms of metric functionals and interpret the result as concretely as possible.

The very last part can in fact often be done, as testified by Theorems 1 and 4 above which include no reference to metric functionals (or horofunctions), and likewise in Theorem 10 of the last section on deep learning. Sometimes, like in complex analysis, Cayley graphs, or maps of cones there is no need to pass to an auxiliary space in order to find an invariant metric for the transformations in question.

Further generalized Lyapunov exponents could perhaps also be defined for higher dimensional diffeomorphisms, using their isometric action on Ebin's space of Riemannian metrics on a fixed compact manifold [16] and an investigation of the metric functionals. Some progress and possibilities are pointed out in Sect. 5.

A different direction, using some of the arguments in [43], was developed by Masai [58], namely for surface bundles over a circle, he showed that for pseudo-Anosov maps the translation length (="top Lyapunov exponent" in the terminology of the present paper) in a certain weak metric equals the 3-dimensional hyperbolic volume of the mapping torus that the map defines.

2 Deep Learning

Deep learning provided Artificial Intelligence (AI) with a long sought-after new tool that moreover exceeded all expectation, as was realized starting from around 2012. The development of deep neural network had begun much earlier and Bengio, Hinton, and LeCun received the 2018 Turing award for these methods. Part of the prize citation stated "By dramatically improving the ability of computers to make sense of the world, deep neural networks are changing not just the field of computing, but nearly every field of science and human endeavor."

The remarkable success of these methods indicates that real-life data tend to have a compositional structure. More precisely, given a learning task, one seeks maps $g_1, g_2 ... g_n$ such that their composition

$$u(n) := g_1 g_2 ... g_n$$

applied to the input data should be close to the desired output (possibly after applying a certain *decision function* f), see Fig. 2. The *depth* n can be several hundred. The maps are often of the form $g_i(x) = \sigma(W_i x + b_i)$ where σ is a fixed nonlinear function, called *activation function*, applied componentwise, W_i is a $d \times d$ matrix, called *weights*, and b_i is a vector in $\mathbb{R}^d$, called *bias vector*. The dimension d is called the *width*. The nonlinearity, inspired by our brains [59, 66], is crucial (for one thing, the composition of affine maps are again affine, and likewise the composition of polynomials is again a polynomial). Some standard choices are, sigmoid/logistic function ($1/(1+e^{-t})$), TanH ($\tanh(t)$), and ReLU (Rectified Linear Unit, $\sigma(t) = \max\{0, t\}$), with different features and advantages. ReLU has been observed to work particularly well, generally better than smooth functions.

In practice, the weights and biases in the neural network are first randomly selected (*initialization*) and then optimized by stochastic gradient descent on a chosen loss function specific to the task (*training*).

Much of the subject of deep learning consists of empirical observations, there is no or little theoretical understanding, as remarked by many authors. According to [71] "A mathematical theory of deep learning would illuminate how they function, allow us to assess the strengths and weaknesses of different network architectures, and lead to major improvements." Which type of layer maps to take, which σ, how

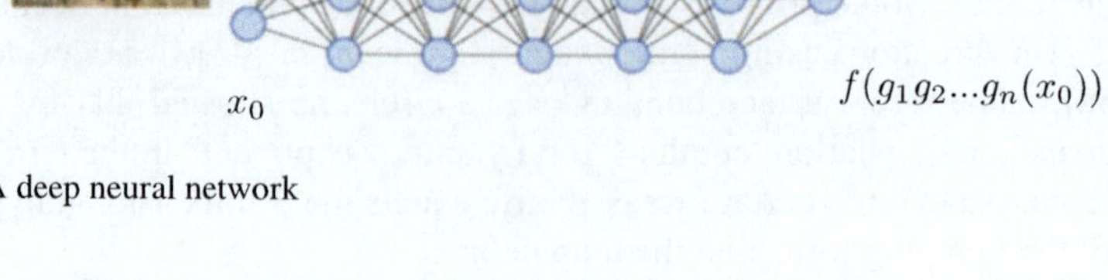

Fig. 2 A deep neural network

many layers n, how many nodes d in each layer, how to best find the parameters, how stable the solution is under random perturbation (such as the *drop-out procedure*) are some of the questions of important practical concern. The need for a theoretical understanding, instead of relying on black-box techniques, is also expressed by practitioners, this lack of theory hinders their work.

One of the remarkable features that is not understood, is why deep neural networks generally seem to mostly avoid the problem of overfitting which is a phenomenon in traditional statistics. The latter typically happens when approximating some data with a polynomial of very high degree, the curves go through all the sample or training data, but inbetween these points of perfect fit it can fluctuate wildly, related to the Runge phenomenon. This is clearly undesirable.

There are several ways random products $u(n) := g_1 g_2 ... g_n$ of noncommuting nonlinear maps appear in deep learning:

1. *Random initialization* see [60] for a review
2. *Drop-out regularization* which in particular is used to verify robustness of the obtained error minimizer, and also a way of training the network [72]
3. *Bayesian learning* [61]
4. Learning that combines taking some maps g_i at random and optimize the remaining one, a procedure with apparently good performance that speeds up the training significantly [4]

The first two concepts are so fundamental in the current state-of-the-art that one encounters them after any couple of first lectures on deep learning. Hanin wrote in [30] "Beyond illuminating the properties of networks at the start of training, the analysis of random neural networks can reveal a great deal about networks after training as well."

Moreover, as Avelin pointed out to me, not only the initialization but also the training (stochastic gradient descent) actually involves a random product of transformations. Thus we see compositional product of random operations appearing in several ways in deep learning. Since the number of layer maps can approach a thousand, it should make limit theorems discussed in this article very relevant, see the last section.

3 Elements of a Metric Functional Analysis

A metric space is a set equipped with a distance function $d(x, y)$ that is semi-positive, symmetric and satisfy the triangle inequality. The author argued in [41] that it is useful to develop parts of metric geometry in analogy with linear functional analysis.

Sometimes various ways of weakening the notion of a metric are useful: pseudo-metrics arise naturally in complex analysis and here we will also allow for asymmetric metrics (as in Thurston's metric). Moreover we will let d possibly to take negative values, useful for topical maps. In other words, we consider weak metrics as defined in the introduction. Weak metrics (but taking only nonnegative values) were in fact already of interest to people like Heinz Hopf in the 1940s, see [65] pointed out in [64].

Note that a symmetrization such as $D(x, y) := \max\{d(x, y), d(y, x)\}$ (or the sum) is nonnegative following from

$$0 = d(x, x) \leq d(x, y) + d(y, x).$$

With the pseudo-metric D one defines a topology on X. In case the separation axiom holds, i.e. that $d(x, y) = 0$ implies $x = y$, then D is a genuine metric.

A map $f : X \to Y$ is *nonexpansive* if

$$d_Y(f(x_1), f(x_2)) \leq d_X(x_1, x_2)$$

for all $x_1, x_2 \in X$. Note that compositions of nonexpansive maps remain nonexpansive. Moreover, note that if we pass to a symmetrization of a weak metric, then f remains nonexpansive.

Let (X, d) be a weak metric space. We will now define the *metric compactification* of X, which will provide a weak topology with compactness properties in the metric setting. (I learnt from Cormac Walsh that this construction works without essential changes to asymmetric metrics, see [1] and [76] which inspired [39], and see also the more recent paper [27].)

Let $F(X, \mathbb{R})$ be the space of continuous functions $X \to \mathbb{R}$ equipped with the topology of pointwise convergence. Given a base point x_0 of the metric space X, let

$$\Phi : X \to \mathrm{F}(X, \mathbb{R})$$

be defined via

$$x \mapsto h_x(\cdot) := d(\cdot, x) - d(x_0, x).$$

Proposition 5 *The map Φ is a well-defined continuous map, and it is injective if d separates points. The closure $\overline{\Phi(X)}$ is compact.*

Proof By the triangle inequality

$$h_x(y) - h_x(z) = d(y, x) - d(z, x) \leq d(y, z)$$

$$-h_x(y) + h_x(z) = -d(y, x) + d(z, x) \leq d(z, y)$$

therefore

$$|h_x(y) - h_x(z)| \leq \max\{d(y, z), d(z, y)\},$$

which in particular implies that h_x is continuous. This inequality clearly passes to the closure. The map Φ is continuous since

$$\left|h_x(z) - h_y(z)\right| = |d(z, x) - d(x_0, x) - d(z, y) + d(x_0, y)| \leq$$

$$2 \max \{d(x, y), d(y, x)\}$$

by the usual triangle inequality and the one in X.

Suppose that d separates points, then given two points x and y, assume that $d(x_0, x) \geq d(x_0, y)$. If $d(x, y) > 0$, then

$$h_x(x) - h_y(x) = -d(x_0, x) - d(x, y) + d(x_0, y) \leq -d(x, y) < 0$$

shows that the two functions are different. In case $d(y, x) > 0$, then

$$h_y(x) - h_x(x) = d(y, x) - d(x_0, y) + d(x_0, x) \geq d(y, x) > 0.$$

These two cases cover all possibilities in view of the remark above about D, and proves the injectivity.

Finally note that by the triangle inequality

$$-d(x_0, y) \leq h_x(y) \leq d(y, x_0).$$

In view of this and the topology of pointwise topology which is the product topology, the Tychonov theorem implies that $\overline{\Phi(X)}$ is compact. □

This proposition is the metric space analog of the Banach–Alaoglu theorem. We call $\overline{X} := \overline{\Phi(X)}$ the *metric compactification of* X and its elements *metric functionals,* which recently has been described concretely in a variety of metric spaces. This development is in parallel to the determination of dual spaces in the beginning of functional analysis a century ago. I reserve the more commonly used word *horofunction* for limits in the topology of uniform convergence on bounded sets (Gromov's choice of topology in [28] considering genuine metric spaces) of h_{x_n} for sequences x_n such that $d(x_0, x_n) \to \infty$.

A Note on the Proof of Theorem 3 in the Weak Metric Case while [43] considered a skew-product extension to the boundary, the [25] paper established a new substantial refinement of the subadditive ergodic theorem [48], which we feel is a refinement of the fundamental theorem of Kingman that has potential for further use. It has indeed already found independent dynamical applications in [10, 37, 78]. In his book [73] computer scientist Szpankowski explains why subadditivity and the subadditive ergodic theorem are fundamental for the analysis of algorithms.

As observed in [39] the noncommutative ergodic theorem works with an asymmetric metric d. Here is the verification that it works even for weak metrics. First, it is of importance that

$$a(n,\omega) := d(x_0, u(n,\omega)x_0)$$

is a subadditive cocycle. This is verified as follows:

$$d(x_0, u(n+m,\omega)x_0) \leq d(x_0, u(n,\omega)x_0) + d(u(n,\omega)x_0, u(n+m,\omega)x_0)$$
$$\leq d(x_0, u(n,\omega)x_0) + d(x_0, u(m, T^n\omega)x_0).$$

Kingman's subadditive ergodic theorem then asserts that

$$\lim_{n\to\infty} \frac{1}{n} d(x, u(n,\omega)x)$$

exists a.e. under the integrability condition

$$\int_\Omega |d(x_0, u(1,\omega)x_0)|\, d\mu < \infty.$$

Note that nothing here depends on the choice of x_0 including the value of the limit therefore written with a general point x. And if we assume that (Ω, μ, T) is an ergodic system then this "top Lyapunov" exponent is deterministic, i.e. essentially constant in ω.

The proof of Theorem 3 in the weak metric setting now follows exactly [25, section 3] except that in that reference the order in which the metric is written is reversed.

A Possible Future Direction It seems plausible that often the limits

$$\lim_{n\to\infty} \frac{1}{n} h(u(n,\omega)x_0)$$

exist a.e. for any metric functional h. Evidence and discussion of this appear in some of my earlier papers, for example [38] (this reference also contains an argument why Theorem 3 in the special case of CAT(0)-spaces is equivalent to geodesic ray approximation [36, 46]). From ray approximation (being of sublinear distance to a geodesic ray) and purely geometric reasons (any two geodesic rays have a well-defined linear rate of asymptotic divergence) all the above limits exist for proper CAT(0)-spaces and Gromov hyperbolic spaces. See also [68] for a recent contribution to this topic. On the other hand, it is not true in general in view of the counterexample [49, p. 272] for $\mathbb{R}^d$ with the ℓ^∞ norm and the metric functionals given for example in [42].

4 Multiplicative Ergodic Theorems for Linear Operators

The need for multiplicative ergodic theorems for operators in infinite dimensions has been expressed in the influential articles [17, 54, 67]. In one approach to the 2D Navier-Stokes equation and related evolution equations, the dynamics takes place in infinite dimensional Hilbert spaces. There has been an increasing interest in results on this topics, for example [8, 9, 23, 53, 57]. González-Tokman wrote in [24] that "An important motivation behind the recent work on multiplicative ergodic theorems is the desire to develop a mathematical theory which is useful for the study of global transport properties of real world dynamical systems, such as oceanic and atmospheric flows. Global features of the ocean flow include large scale structures which are important for the global climate." In a different direction, also leading to multiplicative ergodic theorems in Banach spaces is [12], which deals with difference equations with random delays. Such delays are common in models of biological systems, immune response, epidemiology, and economics (see [12] for references).

Given a bounded linear operator A, submultiplicativity implies that

$$\lim_{n\to\infty} \|A^n\|^{1/n}$$

exists and equals the spectral radius. On the other hand expressions

$$\|A^n v\|^{1/n}$$

as $n \to \infty$ may not converge in infinite dimensions, as is well known, see for example the introduction of [69]. This puts a limitation on the validity of Oseledets theroem for operators. Kingman's theorem takes care of the regularity of the growth of the norm, and Theorem 3 may be the appropriate replacement for the second type of more directional behavior (local spectral theory).

The first infinite dimensional extension of Theorem 1 is Ruelle's theorem [67] for compact operators, and other early results were shown by Mañe and Thieullen [74]. A strengthening of this for the Hilbert-Schmidt class was obtained in [46]:

Theorem 6 ([46]) *Let $u(n, \omega)$ be an ergodic cocycle of $Id + A$ operators where A is Hilbert-Schmidt. Then there is a.s. an operator Λ_ω such that*

$$\frac{1}{n}\left(\sum_i (\log \mu_i(n))^2\right)^{1/2} \to 0$$

as $n \to \infty$, where $\mu_i(n)$ are the eigenvalues of the positive part of $\Lambda_\omega^{-n} u(n, \omega)$.

The uniformity of the convergence implicit in the conclusion, thanks to the metric methods, is noteworthy since this is a much stronger statement in infinite dimensions. In finite dimensions the statement is equivalent to Oseledets' theorem.

This metric approach was recently substantially extended to a von Neumann algebra setting with a finite trace in [9], which used the theorem in [46] together with an intricate analysis, especially of completeness properties, of a space of positive operators admitting a finite trace to get nonpositive curvature.

Let Pos be the space of positive operators on a Hilbert space H. This is a convex cone in the Banach space of symmetric operators, and it has the corresponding Thompson metric:

$$d(p,q)=\sup_{v\in H,\|v\|=1}\left|\log\frac{(qv,v)}{(pv,v)}\right|.$$

Invertible bounded operators g act by isometry on this metric space via $p\mapsto gpg^*$. Unless one restricts to subspaces where there is a finite trace, this is not a CAT(0) space. On the other hand, as noted in particular in [11], the fundamental Segal inequality

$$\|\exp(u+v)\|\leq\|\exp(u/2)\exp(v)\exp(u/2)\|$$

for symmetric operators u and v, can be seen as a weak form of nonpositive curvature more in Busemann's sense (with respect to a selection of geodesics). This means that the exponential map exp:Sym $\rightarrow$ Pos is distance preserving on lines from 0 and otherwise distance-increasing.

In finite dimensions, Lemmens has recently determined $\overline{Pos}$ [50]. In infinite dimensions, the task to describe this compactification remains to be done, with some small steps done in [42] in relation to the invariant subspace problem.

As for ergodic cocycles of invertible bounded linear operators, one has from Theorem 3:

Theorem 7 ([25]) *Let* $v(n,\omega)=A(T^{n-1}\omega)...A(T\omega)A(\omega)$ *be an integrable ergodic cocycle of bounded invertible linear operators of a Hilbert space. Denote the square of the positive part*

$$[v(n,\omega)]:=v(n,\omega)^*v(n,\omega).$$

Then for a.e. ω *there is a metric functional* h_ω *on Pos such that*

$$\lim_{n\to\infty}-\frac{1}{n}h_\omega([v(n,\omega)])=\lim_{n\to\infty}\frac{1}{n}\left\|\log[v(n,\omega)]\right\|.$$

Note that this statement by-passes the limitation mentioned above from local spectral theory. We can deduce an a priori weaker statement as follows, with a bit more information than in [25]. A state is a positive linear functional of norm 1 on certain types of algebras of operators. The space of states is compact in the weak*-topology.

Theorem 8 ([25]) *Let* $v(n, \omega) = A(T^{n-1}\omega)...A(T\omega)A(\omega)$ *be an integrable ergodic cocycle of bounded invertible linear operators of a Hilbert space. Denote the square of the positive part*

$$[v(n, \omega)] := v(n, \omega)^* v(n, \omega).$$

Then for a.e. ω *there is a state* f_ω *on the space of bounded linear operators of the form,*

$$f_\omega(A) = s(A\xi, \xi) + (1 - s)\psi(A),$$

where ξ *is a unit vector,* ψ *is a state on the algebra of all bounded linear operators vanishing on all compact operators and* $0 \leq s \leq 1$*, all depending on* ω*, such that*

$$\lim_{n\to\infty} \frac{1}{n} |f_\omega(\log[v(n, \omega)])| = \lim_{n\to\infty} \frac{1}{n} \|\log[v(n, \omega)]\| .$$

Proof Let $u(n, \omega) := v(n, \omega)^*$ and hence $[v(n, \omega)] = u(n, \omega)I$ is the random orbit in Pos . Therefore, as explained in a previous section, we know that the distance

$$d(I, [v(n, \omega)])$$

is a subadditive cocycle. Moreover, thanks to the exact distance properties of the exponential map recalled above, also the distance inside Sym , which is the operator norm

$$\|\log[v(n, \omega)]\|$$

is subadditive. To see this in detail: by the distance preserving of exp on lines we have

$$d(I, [v(n, \omega)]) = \|\log[v(n, \omega)]\|$$

while

$$\begin{aligned} \|\log[v(n, \omega)] - \log[v(n + m, \omega)]\| &\leq d([v(n, \omega)], [v(n + m, \omega)]) \\ &= d(I, [v(m, T^n\omega)]) \\ &= \left\|\log[v(m, T^n\omega)]\right\| . \end{aligned}$$

Thus by the triangle inequality,

$$\|\log[v(n + m, \omega)]\| \leq \|\log[v(n, \omega)]\| + \left\|\log[v(m, T^n\omega)]\right\| .$$

Let $y_n := \log[v(n, \omega)]$ and $\epsilon_n \searrow 0$. Since y_n is a self-adjoint operator we can find a unit vector ξ_n such that

$$|(y_n\xi_n, \xi_n)| > \|y_n\| - \epsilon_n.$$

Let $f_n(A) = (A\xi_n, \xi_n)$ be the corresponding linear functional, which in other words is a vector state. We assume that the right hand side is strictly positive, that is,

$$\tau := \lim_{n\to\infty} \frac{1}{n} \|\log[v(n, \omega)]\| > 0,$$

otherwise there is nothing to prove. From the main subadditive ergodic result in [25] we have for almost every ω, a sequence $n_i \to \infty$ and sequence $\delta_l \to 0$ such that for every i and every $l \leq n_i$,

$$\|\log[v(n_i, \omega)]\| - \left\|\log[v(n_i - l, T^l\omega)]\right\| \geq (\tau - \delta_l)l.$$

In view of this, for any $l \leq n_i$,

$$\begin{aligned}
\|y_l\| \geq \left|f_{n_i}(y_l)\right| &= \left|f_{n_i}(y_{n_i} + y_l - y_{n_i})\right| \\
&= \left|f_{n_i}(y_{n_i}) - f_{n_i}(y_{n_i} - y_l)\right| \geq \left|f_{n_i}(y_{n_i})\right| - \left|f_{n_i}(y_{n_i} - y_l)\right| \\
\geq \left\|y_{n_i}\right\| - \epsilon_{n_i} - \left\|y_{n_i} - y_l\right\| &\geq \|\log[v(n_i, \omega)]\| - \left\|\log[v(n_i - l, T^l\omega)]\right\| - \epsilon_{n_i} \\
&\geq (\tau - \delta_l)l - \epsilon_{n_i}.
\end{aligned}$$

By weak*-compactness letting $i \to \infty$ there is a state $f = f^\omega$ for which

$$\lim_{l\to\infty} \frac{1}{n} |f(y_l)| = \tau,$$

as desired. By Glimm's theorem [22] this state, being a limit of vector states, must be of the form

$$f(A) = s(A\xi, \xi) + (1 - s)\psi(A),$$

where ξ is a unit vector, ψ a state on the algebra of all bounded linear operators on the Hilbert space vanishing on all compact operators and $0 \leq s \leq 1$. □

Remark When the cocycle is composed of compact operators then s must be 1 and f_ω is a pure vector state, which then provides a result pointing in the direction of Ruelle's theorem.

5 Diffeomorphisms

In the introduction topological Lyapunov exponents for surface homeomorphisms were explained. Thurston's powerful measured foliation theory is presumably difficult to generalize to dimensions greater than two (and so far cannot treat the case of products of random homeomorphisms). The metric perspective can on the other hand more easily be generalized. For example, Ebin's Riemannian manifold of Riemannian metrics on a compact manifold is one possibility [16]. The diffeomorphisms act by isometry. This is the replacement for the Teichmüller spaces. There are two variants, one general and one restricted to Riemannian metrics sharing the same volume form, and we would then consider volume preserving diffeomorphisms. These spaces have nonpositive curvature but not always complete.

Here is a related metric taken from [2], and which perhaps has not been considered before. Let M be a compact submanifold of a finite dimensional vector space equipped with a norm $\|\cdot\|$. Consider the following weak metric (which can be symmetrized if needed) on the set X of distance function functions bi-Lipschitz equivalent to $d_0(x, y) = \|x - y\|$:

$$D(d_1, d_2) = \log \sup_{x \neq y} \frac{d_2(x, y)}{d_1(x, y)}.$$

If $T : M \to M$ is a diffeomorphism, it will preserve D-distances, considered as map $(T^*d)(x, y) := d(Tx, Ty)$ since it just permutes the underlying set M. Note that T^* is an adjoint type of map, it reverses the order of composition (and if it is desired to keep the orientation we could instead use the inverse if it exists).

The following was shown in [2], with one ingredient being the main results of [25]:

Theorem 9 (Existence of a Point with Maximal Stretch [2]) *Let*

$$v(n, \omega) = A(T^{n-1}\omega)...A(T\omega)A(\omega)$$

be an integrable ergodic cocycle of diffeomorphisms of M. Then there is a number λ such that

$$\lim_{n \to \infty} \left(\sup_{x \neq y} \frac{\|v(n, \omega)x - v(n, \omega)y\|}{\|x - y\|} \right)^{1/n} = e^{\lambda}.$$

In the case that $\lambda > 0$ then there exists a point $z \in M$ and a sequence $w_i = (x_i, y_i) \in \{(x, y) \in M \times M : x \neq y\}$ such that $w_i \to (z, z)$ and for any $\epsilon > 0$ there is $N > 0$ such that for all $n > N$

$$\frac{\|v(n, \omega)x_i - v(n, \omega)y_i\|}{\|x_i - y_i\|} \geq e^{(\lambda - \epsilon)n}$$

for all i sufficiently large for a fixed n.

In words, it means that given a cocycle there is a.e. a random point z such that nearby this point the cocycle is stretching at a near maximal rate.

Another possibility of measuring distances in 1-dimensional dynamics is the total variation of the logarithm of the derivative as in [19]. This can be extended to a weak metric on diffeomorphisms groups on compact manifolds M via

$$d(f, g) = \sup_{x \in M} \left| \log \frac{|J_g(x)|}{|J_f(x)|} \right|$$

where $|J_f|$ is the determinant of the Jacobian derivative. See [2] for further definitions of this type and their use.

6 Applications of Metrics and Ergodic Theorems in Deep Learning

In [2] Avelin and I introduced new geometric frameworks to the theory of neural networks, that also enabled the application of the noncommutative ergodic theorem. More specifically, we suggested several metrics on the data set making various choices of layer maps nonexpansive. This includes the most standard choices of activation functions (those mentioned above), and with positive, unitary or invertible features for the weights. Since the composition of nonexpansive maps remains nonexpansive, already this guarantees some regularity and absence of wild fluctuations.

- *In several standard models of neural networks it is possible to find semi-invariant metrics. This may help to explain phenomena that have been observed emprically, such as a certain stability that ensures good generalization as opposed to overfitting.*

In addition, in view of the noncommutative ergodic theorem above, when the layer maps are selected at random and the number of layers is large, their compositions are close to being constant functions in some cases, see for example Theorem 10 below. As Dherin and his colleagues from Google and DeepMind informed us this fits very well their theory aimed at explaining why deep learning generalize well and do not overfit data, instead the trained network has a bias towards simple functions [14]. They measure simplicity with their Geometric Complexity notion inspired by the Dirichlet energy. Constant maps have zero complexity. And as the authors argue in [3] and [14] initializations that deliver near-constant functions are an advantage.

- *When the noncommutative ergodic theorem, Theorem 3, is applicable to the neural network, random initilialization gives near-constant maps. According to [3, 14, 34] this is something observed in practice that may be highly desirable by contributing to the simple nature of the functions that the stochastic gradient method tends to find.*

Thus one can say that from this point of view it seems good to choose a network architecture for which Theorem 3 applies. Such situations are discussed in more detail in [2].

To illustrate the above points, here is a sample corollary of Theorem 3: maps used in a popular layer model called ResNets, see [31], are treated in the following result:

Theorem 10 ([2]) *Let* $X = \mathbb{R}^d$ *with the standard scalar product. Consider the layer maps* $T(x) = W^T\sigma(Wx+b)$, *with b a general vector and W having operator norm at most* 1, *and the activation function being either ReLU, TanH or the sigmoid function. When such layer maps are selected i.i.d. under a finite moment condition, it holds a.s. as* $n \to \infty$ *and any* $x_0 \in X$ *that there is a (random) vector* v *such that*

$$\frac{1}{n}T_1T_2...T_nx_0 \to v.$$

The vector v does not depend on the input x_0 thus the limiting map is a (random) constant function, and in the case of a large, but finite, fixed number n of layers, the composed function should be nearly constant.

The infinite-width limit as been more studied than the infinite-depth case here discussed. For investigations about the case when both the width and he depth go to infinity, see [29, 52].

Acknowledgments This text was written in connection with the conference "New Trends in Lyapunov Exponents" in Lisbon 2022. I thank the organizers and especially Pedro Duarte for the invitation to this very pleasant and stimulating week. I also thank Alex Blumenthal for helpful discussions related to the topics of this paper during this meeting.

The author was supported in part by the Swedish Research Council grant 104651320 and the Swiss NSF grants 200020-200400 and 200021-212864.

References

1. Akian, Marianne; Gaubert, Stéphane; Walsh, Cormac The max-plus Martin boundary. Doc. Math. 14 (2009), 195–240.
2. Avelin, B. Karlsson, A. Deep limits and cut-off phenomena for neural networks, J. Mach. Learn. Res. 23 (191) (2022) 29 pp
3. Barrett, David G.T.; Dherin, Benoit, Implicit Gradient Regularization, ICLR2021
4. Basteri, A.; Trevisan, D. Quantitative Gaussian Approximation of Randomly Initialized Deep Neural Networks, arXiv:2203.07379 [cs.LG]
5. Bellman, Richard, Limit theorems for non-commutative operations. I. Duke Math. J. 21 (1954), 491–500.
6. Benoist, Yves; Quint, Jean-François Random walks on reductive groups. Ergebnisse der Mathematik und ihrer Grenzgebiete. 3. Folge. A Series of Modern Surveys in Mathematics, 62. Springer, Cham, 2016. xi+323 pp.

7. Blath, Jochen; Hermann, Felix; Slowik, Martin, A branching process model for dormancy and seed banks in randomly fluctuating environments. J. Math. Biol. 83 (2021), no. 2, Paper No. 17, 40 pp.
8. Blumenthal, Alex A volume-based approach to the multiplicative ergodic theorem on Banach spaces. Discrete Contin. Dyn. Syst. 36 (2016), no. 5, 2377–2403.
9. Bowen, Lewis; Hayes, Ben; Lin, Yuqing, A multiplicative ergodic theorem for von Neumann algebra valued cocycles, Commun. Math. Phys. (2021), arXiv:2006.13293.
10. Cantat, Serge; Dujardin, Romain, Random dynamics on real and complex projective surfaces, arXiv:2006.04394
11. Corach, G.; Porta, H.; Recht, L. Convexity of the geodesic distance on spaces of positive operators. Illinois J. Math. 38 (1994), no. 1, 87–94.
12. Crauel, H.; Doan, T. S.; Siegmund, S. Difference equations with random delay. J. Difference Equ. Appl. 15 (2009), no. 7, 627–647.
13. Crisanti, A.; Paladin, G.; Vulpiani, A. Products of random matrices in statistical physics. With a foreword by Giorgio Parisi. Springer Series in Solid-State Sciences, 104. Springer-Verlag, Berlin, 1993. xiv+166 pp.
14. Dherin, Benoit; Munn, Michael; Rosca, Mihaela, Barrett, David G.T., Why neural networks find simple solutions: the many regularizers of geometric complexity, NeurIPS (2022)
15. Dyson,F.J. The dynamics of a disordered linear chain. Phys. Rev. 92, (1953) 1331–1338
16. Ebin, David G. On the space of Riemannian metrics. Bull. Amer. Math. Soc. 74 1968 1001–1003.
17. Eckmann, J.-P.; Ruelle, D. Ergodic theory of chaos and strange attractors. Rev. Modern Phys. 57 (1985), no. 3, part 1, 617–656.
18. Erschler, Anna Poisson-Furstenberg boundaries, large-scale geometry and growth of groups. Proceedings of the International Congress of Mathematicians. Volume II, 681–704, Hindustan Book Agency, New Delhi, 2010.
19. Eynard-Bontemps, Hélène, Navas, Andrés, (Arc-)connectedness for the space of smooth $\mathbb{Z}$^d-actions on 1-dimensional manifolds, https://arxiv.org/abs/2103.06940
20. Fathi, A, Laudenbach, F, and Poénaru, V. *Travaux de Thurston sur les surfaces*. Astérisque, 66-67. Société Mathématique de France, Paris, 1979. 284 pp.
21. Furstenberg, Harry Noncommuting random products. Trans. Amer. Math. Soc. 108 1963 377–428.
22. Glimm, James, A Stone-Weierstrass theorem for C$*$-algebras, Ann. of Math. 72 (1960), 216–244.
23. González-Tokman, Cecilia; Quas, Anthony A concise proof of the multiplicative ergodic theorem on Banach spaces. J. Mod. Dyn. 9 (2015), 237–255.
24. González-Tokman, Cecilia Multiplicative ergodic theorems for transfer operators: towards the identification and analysis of coherent structures in non-autonomous dynamical systems. Contributions of Mexican mathematicians abroad in pure and applied mathematics, 31–52, Contemp. Math., 709, Aportaciones Mat., Amer. Math. Soc., Providence, RI, 2018.
25. Gouëzel, S; Karlsson, A. Subadditive and multiplicative ergodic theorems, J. Eur. Math. Soc. 22 (2020), no. 6, 1893–1915.
26. Guivarc'h, Y. Sur la loi des grands nombres et le rayon spectral d'une marche aléatoire. (French) Conference on Random Walks (Kleebach, 1979) (French), pp. 47–98, 3, Astérisque, 74, Soc. Math. France, Paris, 1980.
27. Gutiérrez, Armando W.; Walsh, Cormac; Firm non-expansive mappings in weak metric spaces. Arch. Math. (Basel) 119 (2022), no. 4, 389–400.
28. Gromov, M. Hyperbolic manifolds, groups and actions. Riemann surfaces and related topics: Pro- ceedings of the 1978 Stony Brook Conference (State Univ. New York, Stony Brook, N.Y., 1978), pp. 183–213, Ann. of Math. Stud., 97, Princeton Univ. Press, Princeton, N.J., 1981.
29. Hanin, Boris; Nica, Mihai Products of many large random matrices and gradients in deep neural networks. Comm. Math. Phys. 376 (2020), no. 1, 287–322.
30. Hanin, B. Random Neural Networks in the Infinite Width Limit as Gaussian Processes, arXiv:2107.01562 [math.PR]

31. He, K., Zhang, X., Ren, S., and Sun, J. Deep residual learning for image recognition. 2016 IEEE Conference on Computer Vision and Pattern Recognition (CVPR)
32. Horbez, Camille, The horoboundary of outer space, and growth under random automorphisms. Ann. Sci. Éc. Norm. Supér. (4) 49 (2016), no. 5, 1075–1123.
33. Iasemidis, L.D. Sackellares J.C. REVIEW: Chaos Theory and Epilepsy, The Neuroscientist, (1996) 2 issue 2, page(s): 118–126.
34. Jakub, C. Nica, M. Depth degeneracy in neural networks: Vanishing angles in fully connected ReLU Networks on Initialization, arXiv preprint arXiv:2302.09712
35. Kaimanovich, Vadim A.; Masur, Howard The Poisson boundary of the mapping class group. Invent. Math. 125 (1996), no. 2, 221–264.
36. Kaimanovich, Vadim A. The Poisson formula for groups with hyperbolic properties. Ann. of Math. (2) 152 (2000), no. 3, 659–692.
37. Kalinin, Boris; Sadovskaya, Victoria Periodic approximation of Lyapunov exponents for Banach cocycles. Ergodic Theory Dynam. Systems 39 (2019), no. 3, 689–706.
38. Karlsson, Anders, Linear rate of escape and convergence in directions, In: Proceedings of a Workshop at the Schrödinger Institute, Vienna 2001, (Ed. by V.A. Kaimanovich, in collab. with K. Schmidt, W. Woess) de Gruyter, 2004.
39. Karlsson, Anders, Two extensions of Thurston's spectral theorem for surface diffeomorphisms. Bull. Lond. Math. Soc. 46 (2014), no. 2, 217–226.
40. Karlsson, Anders, Elements of a metric spectral theory, In: Dynamics, Geometry and Number Theory: the Impact of Margulis on Modern Mathematics The University of Chicago Press, 2022
41. Karlsson, Anders, From linear to metric functional analysis. Proc. Natl. Acad. Sci. USA 118 (2021), no. 28, Paper No. e2107069118, 5 pp.
42. Karlsson, Anders, A metric fixed point theorem and some of its applications, preprint 2022.
43. Karlsson, Anders; Ledrappier, François On laws of large numbers for random walks. Ann. Probab. 34 (2006), no. 5, 1693–1706.
44. Karlsson, Anders; Ledrappier, François Linear drift and Poisson boundary for random walks. Pure Appl. Math. Q. 3 (2007), no. 4, Special Issue: In honor of Grigory Margulis. Part 1, 1027–1036.
45. Karlsson, Anders; Ledrappier, François Propriété de Liouville et vitesse de fuite du mouvement brownien. C. R. Math. Acad. Sci. Paris 344 (2007), no. 11, 685–690.
46. Karlsson, Anders; Margulis, Gregory A., A multiplicative ergodic theorem and nonpositively curved spaces. Comm. Math. Phys. 208 (1999), no. 1, 107–123.
47. Karlsson, Anders, Monod, Nicolas, Strong law of large numbers with concave moments, unpublished note 2008
48. Kingman, J. F. C. The ergodic theory of subadditive stochastic processes. J. Roy. Statist. Soc. Ser. B 30 1968 499–510.
49. Kohlberg, Elon; Neyman, Abraham Asymptotic behavior of nonexpansive mappings in normed linear spaces. Israel J. Math. 38 (1981), no. 4, 269–275.
50. Lemmens, Bas, Horofunction compactifications of symmetric cones under Finsler distances, arXiv:2111.12468.
51. Lenzhen, Anna; Rafi, Kasra; Tao, Jing Bounded combinatorics and the Lipschitz metric on Teichmüller space. Geom. Dedicata 159 (2012), 353–371.
52. Li, Mufan Bill; Nica, Mihai; Roy, Daniel M. The Neural Covariance SDE: Shaped Infinite Depth-and-Width Networks at Initialization, NeurIPS 2022
53. Lian, Zeng; Lu, Kening Lyapunov exponents and invariant manifolds for random dynamical systems in a Banach space. Mem. Amer. Math. Soc. 206 (2010), no. 967, vi+106 pp.
54. Lian, Zeng; Young, Lai-Sang Lyapunov exponents, periodic orbits, and horseshoes for semiflows on Hilbert spaces. J. Amer. Math. Soc. 25 (2012), no. 3, 637–665.
55. Maher, Joseph, Tiozzo, Giulio, Random walks on weakly hyperbolic groups, J. Reine Angew. Math. 742 (2018), 187–239.

56. Margulis, G. A. Discrete groups of motions of manifolds of nonpositive curvature. (Russian) Proceedings of the International Congress of Mathematicians (Vancouver, B.C., 1974), Vol. 2, pp. 21–34. Canad. Math. Congress, Montreal, Que., 1975
57. Martirosyan, D.; Nersesyan, V. Multiplicative ergodic theorem for a non-irreducible random dynamical system. J. Differential Equations 268 (2020), no. 7, 3564–3598.
58. Masai, Hidetoshi, Compactification and distance on Teichmüller space via renormalized volume, arXiv:2108.06059.
59. McCulloch, W and W. Pitts. A Logical Calculus of Ideas Immanent in Nervous Activity, Bull. Math. Biophys., 5 (1943) 115–133.
60. Narkhede, M.V.; Bartakke, P.P.; and Sutaone, M.S. A review on weight initialization strategies for neural networks, Artificial intelligence review 55 (2022), no. 1, 291–322.
61. Neal, R.M. Bayesian learning for neural networks, volume 118. Springer Science & Business Media, 2012.
62. Neusser, Klaus, Time–varying rational expectations models. J. Econom. Dynam. Control 107 (2019), 103731, 17 pp.
63. Oseledec, V. I. A multiplicative ergodic theorem. Characteristic Ljapunov, exponents of dynamical systems. (Russian) Trudy Moskov. Mat. Obšč. 19 1968 179–210.
64. Papadopoulos, Athanase; Troyanov, Marc Weak Finsler structures and the Funk weak metric. Math. Proc. Cambridge Philos. Soc. 147 (2009), no. 2, 419–437.
65. Ribeiro, Hugo Sur les espaces à métrique faible. Portugal. Math. 4 (1943), 21–40.
66. Frank Rosenblatt, The perceptron: a probabilistic model for information storage and organization in the brain., Psychological review 65 (1958), no. 6, 386.
67. Ruelle, David Characteristic exponents and invariant manifolds in Hilbert space. Ann. of Math. (2) 115 (1982), no. 2, 243–290.
68. Sampaio, Luís Miguel, Regularity of the drift for random walks in groups acting on Gromov hyperbolic spaces, https://arxiv.org/abs/2109.13060
69. Schaumlöffel, Kay-Uwe, Multiplicative ergodic theorems in infinite dimensions, In: Infinite Dimensional Random Dynamical Systems, Lyapunov Exponents, Springer Lecture Notes in Mathematics, volume 1486, (2006) pp 187–195
70. Schmidt,H. Disordered one-dimensional crystals. Phys.Rev.105,425(1957) 425–441
71. Sejnowski, T.J. The unreasonable effectiveness of deep learning in artificial intelligence, Proc. Nat. Acad. Sci. U.S.A. December 1, 2020 117 (48) 30033–30038.
72. Srivastava, N.; Hinton, G.; Krizhevsky, A.; Sutskever, I. and Salakhutdinov, R.. Dropout: a simple way to prevent neural networks from overfitting. The journal of machine learning research, 15(1):1929–1958, 2014.
73. Szpankowski, Wojciech Average case analysis of algorithms on sequences. With a foreword by Philippe Flajolet. Wiley-Interscience Series in Discrete Mathematics and Optimization. Wiley-Interscience, New York, 2001. xxii+551 pp.
74. Thieullen, P. Fibrés dynamiques asymptotiquement compacts. Exposants de Lyapounov. Entropie. Dimension. Ann. Inst. H. Poincaré Anal. Non Linéaire 4 (1987), no. 1, 49–97
75. Thurston, William P. On the geometry and dynamics of diffeomorphisms of surfaces. Bull. Amer. Math. Soc. (N.S.) 19 (1988), no. 2, 417–431.
76. Walsh, Cormac The horoboundary and isometry group of Thurston's Lipschitz metric. Handbook of Teichmüller theory. Vol. IV, 327–353, IRMA Lect. Math. Theor. Phys., 19, Eur. Math. Soc., Zürich, 2014.
77. Zheng, Tianyi, Asymptotic behaviors of random walks on countable groups. To appear in the Proceedings of ICM 2022.
78. Zou, Rui; Cao, Yongluo Livšic theorems for Banach cocycles: existence and regularity. J. Funct. Anal. 280 (2021), no. 5, Paper No. 108889, 37 pp.

On the Multifractal Formalism of Lyapunov Exponents: A Survey of Recent Results

Reza Mohammadpour

Abstract We survey a collection of results in the multifractal analysis for the topological entropy of the level sets of Lyapunov exponents. We discuss the most recent results in the area as well as the main difficulties in developing a general theory. Due to the nonconformality, the Lyapunov exponents are averages of nonadditive sequences of potentials, and thus one cannot use Birkhoff's ergodic theorem or the classical thermodynamic formalism. The results are formulated in terms of Legendre–Fenchel transforms of topological pressures as well as in terms of restricted variational principles of entropies of invariant measures with given Lyapunov exponents.

Keywords Lyapunov Exponents · Variational principle · Multifractal formalism · Thermodynamic formalism

2020 Mathematics Subject Classification 28A80, 28D20, 37D35, 37H15

1 Introduction

Many important characteristics of dynamical systems are given as local asymptotic quantities, such as Birkhoff averages and Lyapunov exponents, which reveal information about a single point or trajectory. This area of dynamical system is called *Multifractal analysis (formalism).*

Multifractal analysis is a sub-area of thermodynamic formalism devoted to study the complexity of level sets of invariant local quantities. Typical examples of these quantities are Birkhoff averages, Lyapunov exponents, local entropies and pointwise dimensions. The geometry of the level sets is usually complicated and in order to quantify its size or complexity tools such as Hausdorff dimension or topological entropy are used. In this note we will be interested in multifractal analysis of

R. Mohammadpour (✉)
Department of Mathematics, Uppsala University, Uppsala, Sweden
e-mail: reza.mohammadpour@math.uu.se

© The Author(s), under exclusive license to Springer Nature Switzerland AG 2023

J. L. Dias et al. (eds.), *New Trends in Lyapunov Exponents*, CIM Series in Mathematical Sciences, https://doi.org/10.1007/978-3-031-41316-2_6

Birkhoff averages and Lyapunov exponents. In particular, we focus on the recent results of multifractal analysis of Lyapunov exponents.

2 Notation and Basic Definitions

In this section, we recall some basic facts and definitions.

2.1 *Symbolic Dynamics*

Assume that $Q = (q_{ij})$ is a $k \times k$ matrix with $q_{ij} \in \{0, 1\}$. The one sided subshift of finite type associated to the matrix Q is a left shift map $T : \Sigma_Q^+ \to \Sigma_Q^+$ i.e., $T(x_n)_{n\in\mathbb{N}_0} = (x_{n+1})_{n\in\mathbb{N}_0}$, where Σ_Q^+ is the set of sequences

$$\Sigma_Q^+ := \{x = (x_i)_{i\in\mathbb{N}_0} : x_i \in \{1, ..., k\} \text{ and } Q_{x_i,x_{i+1}} = 1 \text{ for all } i \in \mathbb{N}_0\};$$

denote it by (Σ_Q^+, T). Similarly, one defines two sided subshift of finite type $T : \Sigma_Q \to \Sigma_Q$, where

$$\Sigma_Q := \{x = (x_i)_{i\in\mathbb{Z}} : x_i \in \{1, ..., k\} \text{ and } Q_{x_i,x_{i+1}} = 1 \text{ for all } i \in \mathbb{Z}\}.$$

When the matrix Q has entries all equal to 1, we say that is the *full shift*. For simplicity, we denote that $\Sigma_Q^+ = \Sigma^+$ and $\Sigma_Q = \Sigma$.

We equip Σ with the following metric: for $x = (x_i)_{i\in\mathbb{Z}}$, $y = (y_i)_{i\in\mathbb{Z}} \in \Sigma$, set

$$d(x, y) := 2^{-t} \tag{2.1}$$

where t is the largest integer such that $x_i = y_i$ for all $|i| < t$. Equipped with such a metric, Σ becomes a compact metric space and T a hyperbolic homeomorphism of Σ.

We say that $i_0...i_{k-1}$ is an *admissible word* if $Q_{i_n,i_{n+1}} = 1$ for all $0 \leq n \leq k-2$. We denote by $\mathcal{L}$ the collection of admissible words and denote by $\mathcal{L}_n$ the set of *admissible words of length n* that is, a word $x_0, .., x_{n-1}$ with $x_i \in \{1, ..., k\}$ such that $Q_{x_i,x_{i+1}} = 1$. When (Σ, T) is a full shift, $\mathcal{L}$ and $\mathcal{L}_n$ are the set of all words and the set of words of length n, respectively. We also denote by $|I|$ the length of $I \in \mathcal{L}$. One can define *n-th level cylinder* $[I]$ as follows:

$$[I] = [i_0...i_{n-1}] := \{x \in \Sigma : x_j = i_j \ \ \forall\, 0 \leq j \leq n-1\},$$

for any $i_0...i_{n-1} \in \mathcal{L}_n$.

We say that the matrix Q is primitive when there exists $n > 0$ such that all the entries of Q^n are positive. It is well-known that a subshift of a finite type associated with a primitive matrix Q is *topologically mixing*. That means, for every open non-empty $U, V \subset \Sigma$, there is N such that for every $n \geq N$, $T^n(U) \cap V \neq \emptyset$.

In the two-sided dynamics, we define the *local stable set*

$$W^s_{\text{loc}}(x) = \{(y_n)_{n\in\mathbb{Z}} : x_n = y_n \quad \text{for all} \quad n \geq 0\}$$

and the *local unstable set*

$$W^u_{\text{loc}}(x) = \{(y_n)_{n\in\mathbb{Z}} : x_n = y_n \quad \text{for all} \quad n \leq 0\}.$$

Furthermore, the global stable and unstable manifolds of $x \in \Sigma$ are

$$W^s(x) := \left\{y \in \Sigma : T^n y \in W^s_{\text{loc}}(T^n(x)) \text{ for some } n \geq 0\right\},$$

$$W^u(x) := \left\{y \in \Sigma : T^n y \in W^u_{\text{loc}}(T^n(x)) \text{ for some } n \leq 0\right\}.$$

2.2 Multilinear Algebra

We denote by $\sigma_1, ..., \sigma_d$ the singular values of the matrix A, which are the square roots of the eigenvalues of the positive semi definite matrix A^*A listed in decreasing order according to multiplicity.

$\{e_1, \ldots, e_d\}$ is the standard orthogonal basis of $\mathbb{R}^d$ and define

$$\wedge^l \mathbb{R}^d := \text{span}\{e_{i_1} \wedge e_{i_2} \wedge ... \wedge e_{i_l} : 1 \leq i_1 \leq i_2 \leq ... \leq i_l \leq d\}$$

for all $l \in \{1, ..., d\}$ with the convention that $\wedge^0\mathbb{R}^d = \mathbb{R}$. It is called the *l-th exterior power of* $\mathbb{R}^d$.

We are interested in the group of $d \times d$ invertible matrices of real numbers $\text{GL}_d(\mathbb{R})$ that can be seen as a subset of $\mathbb{R}^{d^2}$. This space has a topology induced from $\mathbb{R}^{d^2}$. For $A \in \text{GL}_d(\mathbb{R})$, we define an invertible linear map $A^{\wedge l} : \wedge^l\mathbb{R}^d \to \wedge^l\mathbb{R}^d$ as follows

$$(A^{\wedge l}(e_{i_1} \wedge e_{i_2} \wedge ... \wedge e_{i_l})) = Ae_{i_1} \wedge Ae_{i_2} \wedge ... \wedge Ae_{i_l}.$$

$A^{\wedge l}$ can be represented by a $\binom{d}{l} \times \binom{d}{l}$ whose entries are the $l \times l$ minors of A. It can be also shown that

$$(AB)^{\wedge l} = A^{\wedge l} B^{\wedge l}, \text{ and } \|A^{\wedge l}\| = \sigma_1(A)...\sigma_l(A).$$

2.3 Legendre Transform

Assume that $f : \mathbb{R}^d \to \mathbb{R} \cup \{+\infty\}$ is a convex function that is not identically equal to $-\infty$. The Legendre transform of f is the function f^* of a new variable t, defined by

$$t \mapsto -f^*(-t) := \inf\{f(x) - \langle t, x \rangle : x \in \mathbb{R}^d\}.$$

It is easy to show that f^* is a convex function and not identically equal to $-\infty$. Let f^{**} be the Legendre transform of f^*. Let $x \in \mathbb{R}^d$. Suppose that f is lower semi continuous at x, i.e., $\liminf_{y \to x} f(y) \geq f(x)$. Then $f^{**}(x) = f(x)$.

2.4 Topological Entropy

Assume that (X, d) is a compact metric space and $T : X \to X$ is a continuous transformation. For any $n \in \mathbb{N}$, we define a new metric d_n on X as follows

$$d_n(x, y) = \max\{d(T^k(x), T^k(y)) : k = 0, ..., n-1\}, \tag{2.2}$$

and for any $\epsilon > 0$, one can define *Bowen ball* $B_n(x, \epsilon)$ that is an open ball of radius $\epsilon > 0$ in the metric d_n around x. That is,

$$B_n(x, \epsilon) = \{y \in X : d_n(x, y) < \epsilon\}.$$

Let $Y \subset X$ and $\epsilon > 0$. We say that a countable collection of balls $\mathcal{Y} := \{B_{n_i}(y_i, \epsilon)\}_i$ covers Y if $Y \subset \bigcup_i B_{n_i}(y_i, \epsilon)$. For $\mathcal{Y} = \{B_{n_i}(y_i, \epsilon)\}_i$, put $n(\mathcal{Y}) = \min_i n_i$. Let $s \geq 0$ and define

$$S(Y, s, N, \epsilon) = \inf \sum_i e^{-sn_i},$$

where the infimum is taken over all collections $\mathcal{Y} = \{B_{n_i}(x_i, \epsilon)\}_i$ covering Y such that $n(\mathcal{Y}) \geq N$. The quantity $S(Y, s, N, \epsilon)$ does not decrease with N, so the limit $S(Y, s, N, \epsilon)$ exists; denote

$$S(Y, s, \epsilon) = \lim_{N \to \infty} S(Y, s, N, \epsilon).$$

There is a critical value of the parameter s, which we denote by $h_{\text{top}}(Y, T, \epsilon)$ such that

$$S(Y, s, \epsilon) = \begin{cases} 0, & s > h_{\text{top}}(Y, T, \epsilon), \\ \infty, & s < h_{\text{top}}(Y, T, \epsilon). \end{cases}$$

Since $h_{\text{top}}(Y, T, \epsilon)$ does not decrease with ϵ, the following limit exists,

$$h_{\text{top}}(Y, T) = \lim_{\epsilon \to 0}(Y, T, \epsilon).$$

We call $h_{\text{top}}(Y, T)$ the *topological entropy* of T restricted to Y or the topological entropy of Y (we denote $h_{\text{top}}(T_{|Y})$). We denote $h_{\text{top}}(X, T) = h_{\text{top}}(T)$.

3 Multifractal Analysis of Birkhoff Averages

Let $f : X \to \mathbb{R}$ be a continuous function over a topological dynamical system (X, T). We denote by $\mathcal{M}(X, T)$ the space of all T-invariant Borel probability measures. Also, we denote by $\mathcal{M}_{\text{erg}}(X, T) \subset \mathcal{M}(X, T)$ the subset of ergodic measure. The pressure $P : C(X) \to \mathbb{R}$ can be defined by the following *variational principle* formula:

$$P(f) := \sup_{\mu \in \mathcal{M}(X,T)} \left\{ h_\mu(T) + \int f d\mu \right\}, \tag{3.1}$$

where h_μ is the measure-theoretic entropy. If the supremum is attained, then such measures will be called *equilibrium states*. When $f \equiv 0$, the pressure $P(0)$ is equal to the topological entropy $h_{\text{top}}(X, T)$, which measures the complexity of the system (X, T). By (3.1),

$$h_{\text{top}}(T) = h_{\text{top}}(X, T) = \sup_{\mu \in \mathcal{M}(X,T)} h_\mu(T). \tag{3.2}$$

If the supremum is attained, then such measures will be called *measures of maximal entropy*.

Let $T : X \to X$ be a continuous map on the compact metric space X. Let $f : X \to \mathbb{R}$ be a continuous function. We denote by $S_n f(x) := \sum_{k=0}^{n-1} f(T^k(x))$ the *Birkhoff sum*, and we call

$$\lim_{n \to \infty} \frac{1}{n} S_n f(x), \tag{3.3}$$

the *Birkhoff average*, and μ-almost every $x \in X$, the Birkhoff average is well-defined.

If μ is an ergodic invariant probability measure, then the Birkhoff average converges to $\int f d\mu$ for μ-almost all points, but there are plenty of ergodic invariant measures, for which the limit exists but converges to a different quantity. Furthermore, there are plenty of points which are not generic points for any ergodic measure or even for which the Birkhoff average does not exist. Therefore, one may ask about the size of the set of points

$$E_f(\alpha) = \left\{ x \in X : \frac{1}{n} S_n f(x) \to \alpha \text{ as } n \to \infty \right\},$$

which we call an *α-level set of Birkhoff spectrum*, for a given value α from the set

$$L_f = \left\{ \alpha \in \mathbb{R} : \exists x \in X \text{ and } \lim_{n\to\infty} \frac{1}{n} S_n f(x) = \alpha \right\},$$

which we call the *Birkhoff spectrum*. The size is usually calculated in terms of either topological entropy or Hausdorff dimension.

This type of question was considered by Barreira and Saussol [5]. There is actually quite a large literature on multifractal analysis (or multifractal formalism) which addresses various questions related to this one; see [9].

Let $T : \Sigma \to \Sigma$ be a topologically mixing subshift of finite type. We denote by $C_T(\Sigma) \subset C(\Sigma)$ the family of continuous functions $\varphi : \Sigma \to \mathbb{R}$ for which there exist $\varepsilon > 0$ and $\kappa > 0$ such that

$$\left| \sum_{k=0}^{n-1} \varphi\left(T^k(x)\right) - \sum_{k=0}^{n-1} \varphi\left(T^k(y)\right) \right| < \kappa$$

whenever $d\left(T^k(x), T^k(y)\right) < \varepsilon$ for every $k = 0, \ldots, n-1$.

It is well-known that (see, e.g. [13]) when (Σ, T) is a topologically subshift of finite type and f is a continuous function, then

$$E_f(\alpha) \neq \emptyset \Leftrightarrow \alpha \in \Omega_f := \left\{ \int f d\mu : \mu \in \mathcal{M}(\Sigma, T) \right\}.$$

Theorem 3.1 ([5, Theorem 5]) *Let $T : \Sigma \to \Sigma$ be a topologically mixing subshift of finite type and $f \in C_T(\Sigma)$. Then,*

$$h_{top}(T_{|E_f(\alpha)}) = \sup \left\{ h_\mu(T) : \mu \in \mathcal{M}(\Sigma, T) \text{ such that } \int f d\mu = \alpha \right\},$$

for all $\alpha \in \mathring{\Omega}_f$.

The above formula is called the *restricted variational principle* (compare with (3.2)). For simplicity, we denote $h_{\text{top}}(E_f(\alpha)) := h_{\text{top}}(T_{|E_f(\alpha)})$.

Fan et al. [13] proved the restricted variational principle formula for higher-dimensional multifractal analysis. Let $\Phi : \Sigma \to \mathbb{R}^d$ be a continuous function. That induces a map $\Phi_* : \mathcal{M}(\Sigma, T) \to \mathbb{R}^d$, called the *projection map*, given by

$$\Phi_*(\mu) = \int \Phi d\mu \;\; \forall \mu \in \mathcal{M}(\Sigma, T).$$

Denote $\Omega_\Phi = \{\Phi_*(\mu) : \mu \in \mathcal{M}(\Sigma, T)\}$.

Theorem 3.2 ([13, Theorem A]) *Let $T : \Sigma \to \Sigma$ be a topologically mixing subshift of finite type and let $\Phi : \Sigma \to \mathbb{R}^d$ be a continuous function. Then,*

$$h_{top}(E_\Phi(\alpha)) = \sup \left\{ h_\mu(T) : \mathcal{M}(\Sigma, T) \text{ such that } \Phi_*\mu = \alpha \right\},$$

for all $\alpha \in \Omega_\Phi$.

4 Multifractal Analysis of Lyapunov Exponents

Let $T : X \to X$ be a topological dynamical system. We say that $\Phi = \{\log \phi_n\}_{n\in\mathbb{N}}$ is a *subadditive potential* over a topological dynamical system (X, T) if each ϕ_n is a continuous positive-valued function on X such that

$$0 < \phi_{n+m}(x) \le \phi_n(x)\phi_m(T^n(x)) \quad \forall x \in X, m, n \in \mathbb{N}.$$

Similarly, we call a sequence of continuous functions (potentials) $\Phi = \{\log \phi_n\}_{n\in\mathbb{N}}$ *super-additive* if $-\Phi = \{-\log \phi_n\}_{n\in\mathbb{N}}$ is sub-additive.

Moreover, $\Phi = \{\log \phi_n\}_{n=1}^{\infty}$ is said to be an *almost additive potential* if there exists a constant $C > 0$ such that for any $m, n \in \mathbb{N}$, $x \in X$, we have

$$C^{-1}\phi_n(x)\phi_m(T^n)(x) \le \phi_{n+m}(x) \le C\phi_n(x)\phi_m(T^n(x)).$$

Let $T : \Sigma \to \Sigma$ be a topologically mixing subshift of finite type. We say that a subadditive potential $\Phi := \{\log \phi_n\}_{n=1}^{\infty}$ over (Σ, T) has *bounded distortion:* there exists $C \ge 1$ such that for any $n \in \mathbb{N}$ and $I \in \mathcal{L}_n$, we have

$$C^{-1} \le \frac{\phi_n(x)}{\phi_n(y)} \le C$$

for any $x, y \in [I]$.

4.1 Subadditive Thermodynamic Formalism

Assume that (X, d) is a compact metric space and $T : X \to X$ is continuous. For any $\epsilon > 0$ a set $E \subset X$ is said to be a (n, ϵ)*-separated subset* of X if $d_n(x, y) > \epsilon$ (see (2.2)) for any two different points $x, y \in E$.

Let $\Phi = \{\log \phi_n\}_{n=1}^{\infty}$ be a subadditive potential over a topological dynamical system (X, T). The space X is endowed with a metric d. We define the *topological pressure* of Φ as follows. We define

$$P_n(T, \Phi, \epsilon) = \sup \left\{ \sum_{x \in E} \phi_n(x) : E \text{ is } (n, \epsilon)\text{-separated subset of } X \right\}.$$

Since $P_n(T, \Phi, \epsilon)$ is a decreasing function of ϵ, we define

$$P(T, \Phi, \epsilon) = \limsup_{n \to \infty} \frac{1}{n} \log P_n(T, \Phi, \epsilon),$$

and

$$P(\Phi) = \lim_{\epsilon \to 0} P(T, \Phi, \epsilon). \tag{4.1}$$

We call $P(\Phi)$ the *topological pressure* of Φ, whose existence of the limit is guaranteed from the subadditivity of Φ.

Remark 4.1 Since the subshift of finite type (Σ, T) is expansive, the pressure $P(\Phi)$ may be expressed as follows:

$$P(\Phi) = \limsup_{n \to \infty} \frac{1}{n} \log \sup \left\{ \sum_{x \in E} \phi_n(x) : E \text{ is } (n, 1)\text{-separated subset of } \Sigma \right\};$$

that is, we can compute the pressure by looking at $(n, 1)$-separated sets, and drop the limit in ϵ from the definition of the pressure (4.1).

Cao et al. [8] prove the subadditive variational principle:

$$P(\Phi) = \sup \left\{ h_\mu(T) + \chi(\mu, \Phi) : \mu \in \mathcal{M}(X, T) \right\}, \tag{4.2}$$

where $h_\mu(T)$ is the measure-theoretic entropy and

$$\chi(\mu, \Phi) := \lim_{n \to \infty} \int \frac{\log \phi_n(x)}{n} d\mu(x).$$

An invariant measure $\mu \in \mathcal{M}(X, T)$ achieving the supremum in (4.2) is called an *equilibrium state* of Φ. Moreover, at least one equilibrium state necessarily exists for any subadditive potential Φ if the entropy map $\mu \mapsto h_\mu(T)$ is upper semi-continuous.

4.2 An Example of Subadditive Potentials: Matrix Cocycles

Let $T : X \to X$ be a continuous map on the compact metric space X. We now replace Birkhoff sums by matrix products. That is, given a continuous map $\mathcal{A} : X \to \mathrm{GL}_d(\mathbb{R})$ taking values in the space of $d \times d$ invertible matrices, we consider the products

$$\mathcal{A}^n(x) = \mathcal{A}(T^{n-1}(x))\mathcal{A}(T^{n-2}(x)) \cdots \mathcal{A}(x).$$

The pair $(\mathcal{A}, T)$ is called a *matrix cocycle*. It induces a skew-product $F : X \times \mathbb{R}^d \to X \times \mathbb{R}^d$ as

$$F(x, v) = (T(x), \mathcal{A}(x)v).$$

We say that F is generated by T and $\mathcal{A}$. Observe that $F^n(x, v) = (T^n(x), \mathcal{A}^n(x)v)$ for each $n \geq 1$.

We denote by $\|\mathcal{A}\|$ the Euclidean operator norm of a matrix $\mathcal{A}$, that is submultiplicative i.e.,

$$0 < \|\mathcal{A}^{n+m}(x)\| \leq \|\mathcal{A}^n(x)\| \|\mathcal{A}^m(T^n(x))\| \quad \forall x \in X, m, n \in \mathbb{N},$$

therefore, the potential $\{\log \|\mathcal{A}^n\|\}_{n=1}^{\infty}$ is subadditive. Then, by Kingman's subadditive ergodic theorem, for any $\mu \in \mathcal{M}(X, T)$ and μ-almost every $x \in X$, the following limit, called the *top Lyapunov exponent* at $x \in X$, exists:

$$\chi(x, \mathcal{A}) := \lim_{n\to\infty} \frac{1}{n} \log \|\mathcal{A}^n(x)\|. \tag{4.3}$$

A well-known example of matrix cocycles is *one-step cocycles* which are defined as follows. Assume that $\Sigma = \{1, ..., k\}^{\mathbb{Z}}$ is a symbolic space. Suppose that $T : \Sigma \to \Sigma$ is a shift map, i.e. $T(x_l)_{l\in\mathbb{Z}} = (x_{l+1})_{l\in\mathbb{Z}}$. Given an k-tuple of matrices $\mathbf{A} = (A_1, \ldots, A_k) \in \mathrm{GL}_d(\mathbb{R})^k$, we associate with it the locally constant map $\mathcal{A} : \Sigma \to \mathrm{GL}_d(\mathbb{R})$ given by $\mathcal{A}(x) = A_{x_0}$, that means the matrix cocycle $\mathcal{A}$ depends only on the zero-th symbol x_0 of $(x_l)_{l\in\mathbb{Z}}$. In this case, we say that $(\mathcal{A}, T)$ is a one-step cocycle; when the context is clear, we say that $\mathcal{A}$ is a one-step cocycle. The k-tuple of matrices $\mathbf{A}$ is called the generator of the cocycle $\mathcal{A}$. For any length n word $I = i_0, \ldots, i_{n-1}$, we denote

$$\mathcal{A}_I := A_{i_{n-1}} \ldots A_{i_0}.$$

Therefore, when $(\mathcal{A}, T)$ is a one-step cocycle,

$$\mathcal{A}^n(x) = A_{x_{n-1}} \ldots A_{x_0}.$$

The matrix cocycles generated by the derivative of a diffeomorphism map or a smooth map $T : X \to X$ on a closed Riemannian manifold X and a family of maps $\mathcal{A}(x) := D_x T : T_x X \to T_{T(x)} X$ are called *derivative cocycles*. Moreover, when $T : X \to X$ is an Anosov diffeomorphism (or expanding map), Bowen [4] showed that there exists a symbolic coding of T by a subshift of finite type. Therefore, one can replace the derivative cocycle of a uniformly hyperbolic map by a matrix cocycle over a subshift of finite type. In general, we know much more about one-step cocycles than about the more general derivative cocycles.

Motivated by the study of the multifractal formalism of Birkhoff averages, the level set of Lyapunov exponents of certain special subadditive potentials $\Phi = \{\log \phi_n\}_{n=1}^{\infty}$ on full shifts have been studied in [11, 12, 17], in which $\phi_n(x) = \|\mathcal{A}^n(x)\|$, where $\|\cdot\|$ denotes the operator norm. To be more precise, we are interested in the size of the set of points

$$E(\alpha) = \left\{ x \in X : \lim_{n\to\infty} \frac{1}{n} \log \left\|\mathcal{A}^n(x)\right\| = \alpha \ \text{ as } \ n \to \infty \right\},$$

which we call an *α-level set of the top Lyapunov exponent*, for a given value α from the set

$$L = \left\{ \alpha \in \mathbb{R} : \exists x \in X \ \text{ and } \lim_{n\to\infty} \frac{1}{n} \log \left\|\mathcal{A}^n(x)\right\| = \alpha \right\},$$

which we call *Lyapunov spectrum.*

Let μ be an T-invariant measure. By Oseledets' theorem, there might exist several Lyapunov exponents. We denote by $\chi_1(x, \mathcal{A}) \geq \chi_2(x, \mathcal{A}) \geq \ldots \geq \chi_d(x, \mathcal{A})$ the Lyapunov exponents, counted with multiplicity, of the cocycle $(\mathcal{A}, T)$. We also write $\chi_i(\mu, \mathcal{A}) := \int \chi_i(x, \mathcal{A}) d\mu(x)$ for $i = 1, \ldots, d$. Therefore, one may ask the topological entropy of the $\vec{\alpha}$-level set. For $\vec{\alpha} := (\alpha_1, \ldots, \alpha_d) \in \mathbb{R}^d$, we define the *$\vec{\alpha}$-level set* as follows

$$E(\vec{\alpha}) = \left\{ x \in X : \lim_{n\to\infty} \frac{1}{n} \log \sigma_i(\mathcal{A}^n(x)) = \alpha_i \text{ for } i = 1, 2, \ldots, d \right\},$$

where $\sigma_1, \ldots, \sigma_d$ are singular values, listed in decreasing order according to multiplicity. We also define the *Lyapunov spectrum*

$$\vec{L} = \left\{ \vec{\alpha} \in \mathbb{R}^d : \exists x \in X \text{ such that } \lim_{n\to\infty} \frac{1}{n} \log \sigma_i(\mathcal{A}^n(x)) = \alpha_i \text{ for } i = 1, 2, \ldots, d \right\}.$$

We define *Falconer's singular value function* $\varphi^s(\mathcal{A})$ as follows. Let $k \in \{0, \ldots, d-1\}$ and $k \leq s < k+1$. Then,

$$\varphi^s(\mathcal{A}) = \sigma_1(\mathcal{A}) \cdots \sigma_k(\mathcal{A}) \sigma_{k+1}(\mathcal{A})^{s-k},$$

and if $s \geq d$, then $\varphi^s(\mathcal{A}) = (\det(\mathcal{A}))^{\frac{s}{d}}$.

For $s := (s_1, \cdots, s_d) \in \mathbb{R}^d$, we define the *generalized singular value function* $\psi^{s_1,\ldots,s_d}(\mathcal{A}) : \mathbb{R}^{d\times d} \to [0, \infty)$ as

$$\psi^{s_1,\ldots,s_d}(\mathcal{A}) := \sigma_1(\mathcal{A})^{s_1} \cdots \sigma_d(\mathcal{A})^{s_d} = \left(\prod_{m=1}^{d-1} \|\mathcal{A}^{\wedge m}\|^{s_m - s_{m+1}}\right) \left\|\mathcal{A}^{\wedge d}\right\|^{s_d}.$$

When $s \in [0, d]$, the singular value function $\varphi^s(\mathcal{A}(\cdot))$ coincides with the generalized singular value function $\psi^{s_1,\ldots,s_d}(\mathcal{A}(\cdot))$ where

$$(s_1, \ldots, s_d) = (\underbrace{1, \ldots, 1}_{m \text{ times}}, s - m, 0, \ldots, 0),$$

with $m = \lfloor s \rfloor$. We denote $\psi^s(\mathcal{A}) := \psi^{s_1,\ldots,s_d}(\mathcal{A})$. For any $q \in \mathbb{R}^d$, note that $\psi^q(\mathcal{A})$ is neither submultiplicative nor supermultiplicative. For one-step cocycles, the limsup topological pressure of $\log \psi^q(\mathcal{A})$ can be defined by

$$P^*(\log \psi^q(\mathcal{A})) := \limsup_{n\to\infty} \frac{1}{n} \log s_n(q), \qquad \forall q \in \mathbb{R}^d,$$

where $s_n(q) := \sum_{I\in\mathcal{L}_n} \psi^q(\mathcal{A}_I)$. When the limit exists, we denote the topological pressure by $P(\log \psi^q(\mathcal{A}))$.

4.3 Multifractal Analysis of the Top Lyapunov Exponent

Given a matrix $A \in \mathrm{GL}_2^+(\mathbb{R})$, define $f_A : \mathbb{P}^1 \to \mathbb{P}^1$ by

$$f_A(v) = \frac{Av}{\|Av\|}.$$

We denote by $\langle \mathbf{A} \rangle$ the semigroup generated by $\mathbf{A}$. An element $A \in \mathrm{SL}_2(\mathbb{R})$ is elliptic if the absolute value of its trace is strictly less than 2 ; in such a case the matrix A is conjugate to a rotation by some angle. An element $A \in \mathrm{SL}_2(\mathbb{R})$ is hyperbolic if the absolute value of its trace is strictly larger than 2, which is equivalent to the fact that the matrix A has one eigenvalue with absolute value bigger than one and one smaller than one. The set $\mathfrak{E}_N$ of elliptic cocycles is the set of cocycles $\mathcal{A}$ generated by $\mathbf{A} \in \mathrm{SL}_2(\mathbb{R})^N$ such that $\langle \mathbf{A} \rangle$ contains an elliptic element.

Let us denote by $\mathfrak{H}_1 \subset \mathrm{SL}_2(\mathbb{R})$ the subset of hyperbolic matrices and by $\mathfrak{I}_1 \subset \mathrm{SL}_2(\mathbb{R})$ the one of "irrational rotations"

$$\mathfrak{I}_1 = \left\{ \begin{pmatrix} \cos 2\pi\theta & \sin 2\pi\theta \\ -\sin 2\pi\theta & \cos 2\pi\theta \end{pmatrix} : \theta \in [0, 1), \theta \notin \mathbb{Q} \right\}.$$

Now let

$$\mathfrak{E}_{N,\,\mathrm{shyp}} = \bigcup_{A\in\mathfrak{H}_1, B\in\mathfrak{I}_1} \{\mathcal{A} \in \mathfrak{E}_N \text{ generated by } \mathbf{A} : \text{there exist } A', B' \in \langle\mathbf{A}\rangle, \text{ and}$$

$$C \in \mathrm{SL}(2,\mathbb{R}) \text{ so that } f_{C^{-1}A'C}, f_{C^{-1}B'C} \text{ are } \varepsilon\text{-close to } f_A, f_B, \text{ respectively} \}.$$

Theorem 4.2 ([11, Theorem B]) *Assume that a one-step cocycle* $\mathcal{A} \in \mathfrak{E}_{N,\,shyp}$. *Then, for every* $N \geq 2$ *the set* $\mathfrak{E}_{N,shyp}$ *is open and dense in* $\mathfrak{E}_N$ *and there are numbers* $0 < \alpha_+ < \alpha_{\max}$ *such that the map* $\alpha \mapsto h_{top}\ (E(\alpha))$ *is continuous and concave on* $[0, \alpha_{\max}]$, *having a unique maximum at* α_+*and*

$$h_{top}\ (E\ (\alpha_+)) = \log N,$$

we have $0 < h_{top}\ (E(0)) < \log N$, *and for every* $\alpha \in (0, \alpha_{\max}]$ *we have*

$$h_{top}\ (E(\alpha)) = \sup\left\{h_\mu(T) : \mu \in \mathcal{M}_{\mathrm{erg}}(\Sigma^+, T), \chi_1(\mu, \mathcal{A}) = \alpha\right\}.$$

For one-step cocycles, Feng [12] proved that the entropy spectrum of the top Lyapunov exponent is equal to the Legendre-transform of the topological pressure. Assume that $(A_1, \ldots, A_k) \in \mathrm{GL}_d(\mathbb{R})^k$ generates a one-step cocycle $\mathcal{A} : \Sigma \to \mathrm{GL}_d(\mathbb{R})$. We say that $\mathcal{A}$ is irreducible if it does not exist a proper subspace $V \subset \mathbb{R}^d$ such that $A_i V \subset V$ for $i = 1, \ldots, k$.

Theorem 4.3 *Assume that* $(A_1, \ldots, A_k) \in GL_d(\mathbb{R})^k$ *generates a one-step cocycle* $\mathcal{A} : \Sigma \to GL_d(\mathbb{R})$. *Suppose that* $\mathcal{A} : \Sigma \to GL_d(\mathbb{R})$ *is irreducible. Then,*

$$h_{top}(E(\alpha)) = \inf_{q\in\mathbb{R}}\{P(\log \psi^q(\mathcal{A})) - \alpha q\},$$

for $\alpha \in L$.

Proof Take a slight modification of the proof of [12, Theorem 1.1]. □

5 Multifractal Analysis of All Lyapunov Exponents

In this section, we discuss results that calculate the entropy spectrum of all Lyapunov exponents.

5.1 *Multifractal Analysis of Vectors of Lyapunov Exponents for Dominated Matrix Cocycles*

5.1.1 Domination

Let V be a continuous d-dimensional vector bundle over a compact Hausdorff space X, and of finite dimension d. Denote by V_x the fibre over $x \in X$. Fix on V a (continuous) inner product $\langle \cdot, \cdot \rangle$, and let $\| \cdot \|$ be the induced norm. If $B : V_x \to V_y$ is a linear map, its norm $\|B\|$ and co-norm $\mathrm{m}(B)$ are defined respectively as the supremum and the infimum of $\|Bv\|$ over the unit vectors $v \in V_x$.

Let $\mathbb{A} : V \to V$ be a vector bundle automorphism, fibering over a homeomorphism $T : X \to X$. We also call $\mathbb{A}$ a cocycle. For each $x \in X$ and $n \in \mathbb{Z}$, we let $\mathcal{A}^n(x)$ be the restriction of $\mathbb{A}^n$ to the fibre V_x; it is a linear map from V_x to $V_{T^n(x)}$. We write $\mathcal{A}(x) = \mathcal{A}^1(x)$. In the case where the vector bundle is trivial, i.e., $V = X \times \mathbb{R}^d$, we can regard $\mathcal{A}$ as a map $\mathcal{A} : X \to \mathrm{GL}_d(\mathbb{R})$; by abuse of notation we write $\mathbb{A} = (\mathcal{A}, T)$.

A splitting $E \oplus F$ for the bundle V is a continuous family of splittings $E_x \oplus F_x = V_x$, where we require the dimensions of E_x and F_x to be constant. The dimension of E is called the *index* of the splitting. The splitting is invariant (with respect to $\mathbb{A}$) if $\mathcal{A}(x) \cdot E_x = E_{T(x)}$, $\mathcal{A}(x) \cdot F_x = F_{T(x)}$.

A splitting $V = E \oplus F$ is called *dominated* for $\mathbb{A}$ if it is invariant and there are constants $C > 0$ and $0 < \tau < 1$ such that

$$\frac{\|\mathcal{A}^n(x) \mid F_x\|}{\mathrm{m}\left(\mathcal{A}^n(x) \mid E_x\right)} < C\tau^n \quad \text{for every } x \subset X \text{ and every } n \geq 0.$$

It is also said that E dominates F.

In the matrix cocycles setting, we are interested in bundles of the form $X \times \mathbb{R}^d$, where the matrix cocycles are generated by $(\mathcal{A}, T)$. It was showed [2] that a cocycle admits a dominated splitting $V \oplus W$ with $\dim W = i$ if and only if when $n \to \infty$, the ratio between the i-th and $(i + 1)$-th singular values of the matrices of the n-th iterate increase uniformly exponentially.

Definition 5.1 We say that a matrix cocycle $\mathcal{A}$ is *dominated with index i* if there exist constants $C > 1$, $0 < \tau < 1$ such that

$$\frac{\sigma_{i+1}(\mathcal{A}^n(x))}{\sigma_i(\mathcal{A}^n(x))} \leq C\tau^n, \quad \forall n \in \mathbb{N}, x \in X.$$

We say that the cocycle $\mathcal{A}$ is *dominated* if $\mathcal{A}$ is dominated with index i for all $i \in \{1, \ldots, d-1\}$.

Let $\mathbf{A}$ be a compact set in $\mathrm{GL}_d(\mathbb{R})$. We say that $\mathbf{A}$ is *dominated* of index i iff there exist $C > 0$ and $0 < \tau < 1$ such that for any finite sequence $A_1, \ldots, A_N$ in $\mathbf{A}$

we have

$$\frac{\sigma_{i+1}(A_1 \cdots A_N)}{\sigma_i(A_1 \cdots A_N)} < C\tau^N.$$

We say that **A** is dominated iff it is dominated of index i for each $i \in \{1, \ldots, d-1\}$. A one step cocycle $\mathcal{A}$ generated by **A** is dominated if **A** is dominated.

Remark 5.2 According to the multilinear algebra properties, $\mathcal{A}$ is dominated with index k if and only if $\mathcal{A}^{\wedge k}$ is dominated with index 1. Therefore, the cocycle $\mathcal{A}$ is dominated if and only if $\mathcal{A}^{\wedge k}$ is dominated with index k for any $k \in \{1, \ldots, d-1\}$.

5.1.2 Cone-Criterion

Let $f : M \to M$ be a C^r $(r \geq 1)$ diffeomorphism on a compact d-dimensional Riemannian manifold. For $K \subset M$, one denotes $T_K M = \bigcup_{x \in K} T_x M \subset TM$ with the topology induced by the inclusion. If $E_x \subset T_x M$ is a subbundle of $T_x M$ one denotes $Df|_{E_x}$ to the linear map $Df|_{E_x} : E_x \to D_x f(E_x)$, which is the restriction of $D_x f$ to E_x.

We adopt the convention that if V is a vector space, a cone C in V is a subset such that there exists non-degenerate quadratic form Q_C such that

$$C = \{v \in V : Q_C(v) \geq 0\}.$$

The interior of a cone is int $C = \{v \in V : Q_C(v) > 0\} \cup \{0\}$.

A *cone-field* $\mathcal{C}$ on $K \subset M$ is given by:

- a (not necessarily invariant) splitting $T_K M = E \oplus F$ into continuous subbundles whose fibers have dimension d_- and d_+, respectively,
- a continuous family of Riemannian norms $\|\cdot\|$ defined on $T_K M$ (not necessarily the ones given by the underlying Riemannian metric).

In this setting, for $x \in K$, one associates the cone

$$\mathcal{C}_x = \{v = v_E \oplus v_F \in T_x M, \|v_F\| \geq \|v_E\|\}.$$

The dimension $\dim \mathcal{C}$ of the cone-field $\mathcal{C}$ is the dimension d_+ of the bundle F.

We say that the cone-field $\mathcal{C}$ defined in K is Df-invariant if there exists $N > 0$ such that for every $x \in K \cap \cdots \cap f^{-N}(K)$ one has that

$$D_x f^N(\mathcal{C}_x) \subset \operatorname{int} \mathcal{C}_{f^N(x)}.$$

Domination can be characterized in terms of the existence of invariant cone fields for derivative cocycles ([10, Theorem 2.6]).

Let Λ be a repeller of a C^1 map $f : \mathbb{R}^2 \to \mathbb{R}^2$. This means that Λ is compact, f-invariant $\left(f^{-1}\Lambda = \Lambda\right)$, and that f is expanding on Λ, i.e., there exist $c > 0$ and $\beta > 1$ such that

$$\left\|D_x f^n v\right\| \geq c\beta^n \|v\| \text{ for every } x \in \Lambda, n \in \mathbb{N}, \text{ and } v \in T_x\mathbb{R}^2.$$

We also assume that there is an open set $U \supset \Lambda$ such that $\Lambda = \bigcap_{n\in\mathbb{N}} f^n U$, and that $f_{|\Lambda}$ is topologically mixing. Barreira and Gelfert [1] used a subadditive topological pressure to characterize the topological entropy of each level set. They show that the entropy spectrum of Lyapunov exponents is equal to the Legendre transform of the topological pressure (see Sect. 2.3).

Theorem 5.3 ([1, Theorem 1]) *Let Λ be a repeller of a $C^{1+\alpha}$ map $f : \mathbb{R}^2 \to \mathbb{R}^2$ such that:*

- $\{\log Df^n\}_{n=1}^{\infty}$ *has bounded distortion on Λ;*
- *there is a cone-field $\mathcal{C} = \{\mathcal{C}_x\}_{x\in\Lambda}$.*

Then for each $q \in \mathbb{R}^2$ and each $\vec{\alpha} \in \nabla P(\log \psi^q(Df))$,

$$h_{\text{top}}(E(\vec{\alpha})) = \inf_{q\in\mathbb{R}^2}\{P(\log \psi^q(Df)) - \langle q, \vec{\alpha}\rangle\}.$$

A crucial technique in their work is as follows. Barreira and Gelfert [1] considered a C^1 local diffeomorphism $f : \mathbb{R}^2 \to \mathbb{R}^2$, and a compact f-invariant set $\Lambda \subset \mathbb{R}^2$. They showed that if there is a dominated splitting $T_\Lambda\mathbb{R}^2 := E_x \oplus F_x$, then there exists $C \geq 1$ such that for every $x \in \Lambda$ and $n, m \in \mathbb{N}$ we have

$$C^{-1}\sigma_i\left(D_x f^n\right)\sigma_i\left(D_{f^n x} f^m\right) \leq \sigma_i\left(D_x f^{n+m}\right) \leq \sigma_i\left(D_x f^n\right)\sigma_i\left(D_{f^n x} f^m\right).$$

The author [17] extends their result to any dominated cocycle in arbitrary dimensions. As we said, domination can be characterized in terms of the existence of invariant cone fields.

Theorem 5.4 ([17, Proposition 5.8]) *Let X be a compact metric space, and let $\mathcal{A} : X \to GL_d(\mathbb{R})$ be a matrix cocycle over a homeomorphism (X, T). Assume that $\mathcal{A}$ is dominated with index 1 on X. Then, there exists $\kappa > 0$ such that for every $m, n > 0$ and for every $x \in X$ we have*

$$||\mathcal{A}^{m+n}(x)|| \geq \kappa||\mathcal{A}^m(x)|| \cdot ||\mathcal{A}^n(T^m(x))||.$$

Feng and Huang [14] generalized Theorem 5.3 to almost additive potentials over a topologically mixing subshifts of finite type. In particular, they showed the two additional assumptions in Theorem 5.3 can be removed.

Theorem 5.5 ([14, Theorem 5.2]) *Let (Σ, T) be a topologically mixing subshift of finite type, and $\Phi_i := \{\log \phi_n^i\}_{n\in\mathbb{N}}$ $(i = 1, \ldots, d)$ be an almost additive potential* *on* Σ. *Assume that* $\vec{\Omega}$ *is the range of the map from* $\mathcal{M}(\Sigma, T)$ *to* $\mathbb{R}^d$

$$\mu \mapsto (\chi(\mu, \Phi^1), \chi(\mu, \Phi^2), ..., \chi(\mu, \Phi^d)).$$

Then

$$\begin{aligned} h_{top}(E(\vec{\alpha})) &= \inf_{q \in \mathbb{R}^d} \left\{ P\left(\sum_{i=1}^{d} q_i \log \Phi_i\right) - \langle q, \vec{\alpha} \rangle \right\} \\ &= \sup\{h_\mu(T) : \mu \in \mathcal{M}(\Sigma, T), (\chi(\mu, \Phi_1), \ldots, \chi(\mu, \Phi_d)) = \vec{\alpha}\} \end{aligned}$$

for any $\vec{\alpha} \in \vec{\Omega}$.

By using Theorems 5.4 and 5.5, one can extend Theorem 5.3 to the higher dimensional case.

Corollary 5.6 *Assume that* $\mathcal{A} : \Sigma \to GL_d(\mathbb{R})$ *is dominated over a topologically mixing subshift of finite type* (Σ, T). *Assume that* $\vec{\Omega}_{\mathcal{A}}$ *is the range of the map from* $\mathcal{M}(X, T)$ *to* $\mathbb{R}^d$

$$\mu \mapsto (\chi_1(\mu, \mathcal{A}), \chi_2(\mu, \mathcal{A}), ..., \chi_d(\mu, \mathcal{A})).$$

Then,

$$\begin{aligned} h_{top}(E(\vec{\alpha})) &= \inf_{q \in \mathbb{R}^d} \left\{ P(\log \psi^q(\mathcal{A})) - \langle q, \vec{\alpha} \rangle \right\} \\ &= \sup\{h_\mu(T) : \mu \in \mathcal{M}(\Sigma, T), (\chi_1(\mu, \mathcal{A}), \ldots, \chi_d(\mu, \mathcal{A})) = \vec{\alpha}\} \end{aligned}$$

for any $\vec{\alpha} \in \vec{\Omega}_{\mathcal{A}}$.

Proof By Theorem 5.4, $\{\log \sigma_i(\mathcal{A}^n)\}_{n=1}^{\infty}$ is almost additive for all $i = 1, \ldots, d$ (see Remark 5.2). Then, the proof follows from Theorem 5.5. □

5.2 *Multifractal Formalism of Vectors of Lyapunov Exponents for Generic Matrix Cocycles*

Theorem 5.5 showed that the restricted variational principle formula for a vector of Lyapunov exponents holds for almost additive potentials. Note that almost additivity condition holds only for a restrictive family of matrices. For instance, Bárány, Käenmäki and Morris [3, Corollary 2.5] showed that this condition for planar matrix tuples is equivalent to the domination. In this final subsection, we discuss multifractal analysis of all Lyapunov exponents for typical cocycles, which is open and dense.

5.2.1 Typical Cocycles

We say that $\mathcal{A} : \Sigma \to \mathrm{GL}_d(\mathbb{R})$ is an α-Hölder continuous function, if there exists $C > 0$ such that

$$\|\mathcal{A}(x) - \mathcal{A}(y)\| \le C d(x, y)^{\alpha} \ \ \forall x, y \in \Sigma. \tag{5.1}$$

Definition 5.7 A *local stable holonomy* for the matrix cocycle $(\mathcal{A}, T)$ is a family of matrices $H^s_{y\leftarrow x} \in \mathrm{GL}_d(\mathbb{R})$ defined for all $x \in \Sigma$ with $y \in W^s_{\mathrm{loc}}(x)$ such that

(a) $H^s_{x\leftarrow x} = Id$ and $H^s_{z\leftarrow y} \circ H^s_{y\leftarrow x} = H^s_{z\leftarrow x}$ for any $z, y \in W^s_{\mathrm{loc}}(x)$.
(b) $\mathcal{A}(y) \circ H^s_{y\leftarrow x} = H^s_{T(y)\leftarrow T(x)} \circ \mathcal{A}(x)$.
(c) $(x, y, v) \mapsto H^s_{y\leftarrow x}(v)$ is continuous.

Moreover, if $y \in W^u_{\mathrm{loc}}(x)$, then similarly one defines $H^u_{y\leftarrow x}$ with analogous properties.

According to (b) in the above definition, one can extend the definition to the global stable holonomy $H^s_{y\leftarrow x}$ for $y \in W^s(x)$ not necessarily in $W^s_{\mathrm{loc}}(x)$:

$$H^s_{y\leftarrow x} = \mathcal{A}^n(y)^{-1} \circ H^s_{T^n(y)\leftarrow T^n(x)} \circ \mathcal{A}^n(x), \tag{5.2}$$

where $n \in \mathbb{N}$ is large enough such that $T^n(y) \in W^s_{\mathrm{loc}}(T^n(x))$. One can extend the definition of the global unstable holonomy similarly.

Definition 5.8 An α-Hölder continuous function $\mathcal{A}$ is *fiber bunched* if for any $x \in \Sigma$,

$$\|\mathcal{A}(x)\|\|\mathcal{A}(x)^{-1}\| < 2^{\alpha} \tag{5.3}$$

(see (2.1)).

We say that the matrix cocycle $(\mathcal{A}, T)$ is fiber-bunched if its generator $\mathcal{A}$ is fiber-bunched. We denote by $H^{\alpha}_b(\Sigma, \mathrm{GL}_d(\mathbb{R}))$ the space of α-Hölder and fiber-bunched $\mathrm{GL}_d(\mathbb{R})$-valued functions.

The Hölder continuity and the fiber bunched assumption on $\mathcal{A} \in H^{\alpha}_b(\Sigma, \mathrm{GL}_d(\mathbb{R}))$ implies the convergence of the *canonical holonomy* $H^{s/u}$ (see [15]). That means, for any $y \in W^{s/u}_{\mathrm{loc}}(x)$,

$$H^s_{y\leftarrow x} := \lim_{n\to\infty} \mathcal{A}^n(y)^{-1}\mathcal{A}^n(x) \ \text{ and } \ H^u_{y\leftarrow x} := \lim_{n\to-\infty} \mathcal{A}^n(y)^{-1}\mathcal{A}^n(x).$$

Moreover, the canonical holonomies vary α-Hölder continuously (see [15]), i.e., there exists $C > 0$ such that for $y \in W^{s/u}_{\mathrm{loc}}(x)$,

$$\|H^{s/u}_{y\leftarrow x} - I\| \le C d(x, y)^{\alpha}. \tag{5.4}$$

In this paper, we will always work with the canonical holonomies. Note that the canonical holonomies always exist for one-step cocycles; see [7, Proposition 1.2] and [17, Remark 1].

Let $T : \Sigma \to \Sigma$ be a topologically mixing subshift of finite type. Suppose that $p \in \Sigma$ is a periodic point of T, we say $p \neq z \in \Sigma$ is a *homoclinic point* associated to p if it is the intersection of the stable and unstable manifold of p. That is, $z \in W^s(p) \cap W^u(p)$. We denote the set of all homoclinic points of p by $\mathcal{H}(p)$. Then, we define the *holonomy loop*

$$W_p^z := H^s_{p\leftarrow z} \circ H^u_{z\leftarrow p}.$$

Up to replacing z by some backward iterate, we may suppose that $z \in W^u_{\text{loc}}(p)$ and $T^n(z) \in W^s_{\text{loc}}(p)$ for some $n \geq 1$, which may be taken as a multiple of the period of p. Then, by the analogue of (5.2) for stable holonomies,

$$W_p^z = \mathcal{A}^{-n}(p) \circ H^s_{p\leftarrow T^n(z)} \circ \mathcal{A}^n(z) \circ H^u_{z\leftarrow p}.$$

Definition 5.9 Suppose that $\mathcal{A} : \Sigma \to \text{GL}_d(\mathbb{R})$ is a one-step cocycle or that belongs to $H^\alpha_b(\Sigma, \text{GL}_d(\mathbb{R}))$. We say that $\mathcal{A}$ is *1-typical* if there exist a periodic point p and a homoclinic point z associated to p such that:

(i) The eigenvalues of $\mathcal{A}^{per(p)}(p)$ have multiplicity 1 and distinct absolute values,
(ii) We denote by $\{v_1, \ldots, v_d\}$ the eigenvectors of $\mathcal{A}^{per(p)}(p)$, for any $I, J \subset \{1, \ldots, d\}$ with $|I| + |J| \leq d$, the set of vectors

$$\left\{W_p^z(v_i) : i \in I\right\} \cup \left\{v_j : j \in J\right\}$$

is linearly independent.

We say $\mathcal{A}$ is *typical* if $\mathcal{A}^{\wedge t}$ is 1-typical with respect to the same typical pair (p, z) for all $1 \leq t \leq d-1$.

In the above definition, the only role of the fiber-bunching condition is to assure the convergence of the canonical holonomies. Actually, one can define 1-typicality for matrix cocycles that are not necessarily fiber-bunched, but they still admit canonical holonomies. For example, although the exterior product cocycles $\mathcal{A}^{\wedge t}$ might not necessarily be fiber-bunched, they still admit the canonical holonomies; such examples one can find among one-step cocycles. So one may still consider the 1-typicality assumption on the exterior product cocycles as we did in Definition 5.9.

Remark 5.10 For simplicity, we will always let p be a fixed point by passing to the power $\mathcal{A}^{\text{per}(p)}$ if necessary (because powers of typical cocycles are typical). Moreover, for any homoclinic point $z \in \mathcal{H}(p)$, $T^n(z)$ is a homoclinic point of p for any $n \in \mathbb{Z}$. That implies that if $z \in \mathcal{H}(p)$ satisfies (ii), then so does any point $T^n(z) \in \mathcal{H}(p)$ in its orbit too. Therefore, we can replace z by any point in its orbit without destroying (ii).

Above definition for typical cocycles is a slightly stronger than typical cocycles which were first introduced by Bonatti and Viana [7]; they only require 1-typicality of $\mathcal{A}^{\wedge t}$ for $1 \leq t \leq d/2$, and they do not ask the typical pair (p, z) to be the same pair over different t. Even though our version of typicality is slightly stronger, one can perturb their techniques to show that the set of typical cocycles is open and dense. The author [17] extended Theorem 4.3 to typical cocycles.

Theorem 5.11 ([17, Theorem B]) *Assume that $T : \Sigma \to \Sigma$ is a topologically mixing subshift of finite type. Suppose that $\mathcal{A} : \Sigma \to GL_d(\mathbb{R})$ is a typical cocycle. Assume that $\Omega_{\mathcal{A}}$ is the range of the map from $\mathcal{M}(\Sigma, T)$ to $\mathbb{R}$*

$$\mu \mapsto \chi_1(\mu, \mathcal{A}).$$

Then,

$$\begin{aligned} h_{\text{top}}(E(\alpha)) &= \sup\{h_\mu(T) : \mu \in \mathcal{M}(\Sigma, T), \chi_1(\mu, \mathcal{A}) = \alpha\} \\ &= \inf_{q \in \mathbb{R}} \{P(\psi^q(\mathcal{A})) - \alpha.q\} \ \ \forall \alpha \in \mathring{\Omega}_{\mathcal{A}}. \end{aligned}$$

In the following result, we study the size of the level sets of all Lyapunov exponents. We showed that the topological entropy of the level sets of all Lyapunov exponents for typical cocycles is equal to the Legendre transform of topological pressure of the generalized singular value function.

Theorem 5.12 ([16, Theorem A]) *Assume that $(A_1, \ldots, A_k) \in GL_d(\mathbb{R})^k$ generates a one-step cocycle $\mathcal{A} : \Sigma \to GL_d(\mathbb{R})$. Let $\mathcal{A} : \Sigma \to GL_d(\mathbb{R})$ be a typical cocycle. Then*

$$h_{\text{top}}(E(\vec{\alpha})) = \inf_{q \in \mathbb{R}^d} \left\{P\left(\log \psi^q(\mathcal{A})\right) - \langle q, \vec{\alpha}\rangle\right\}$$

for all $\vec{\alpha} \in \mathring{\vec{L}}$.

In the following result, we establish a variational relation between the Legendre transform of topological pressure of the generalized singular value function and measure-theoretic entropies.

Theorem 5.13 ([18, Theorem 1.1]) *Assume that $(A_1, \ldots, A_k) \in GL_d(\mathbb{R})^k$ generates a one-step cocycle $\mathcal{A} : \Sigma \to GL_d(\mathbb{R})$. Suppose that $\mathcal{A} : \Sigma \to GL_d(\mathbb{R})$ is a typical cocycle. Assume that $\vec{\Omega}_{\mathcal{A}}$ is the range of the map from $\mathcal{M}(\Sigma, T)$ to $\mathbb{R}^d$*

$$\mu \mapsto (\chi_1(\mu, \mathcal{A}), \chi_2(\mu, \mathcal{A}), \ldots, \chi_d(\mu, \mathcal{A})).$$

Then,

$$\sup\Big\{h_\mu(T) : \mu \in \mathcal{M}(\Sigma, T), \chi_i(\mu, \mathcal{A}) = \alpha_i \text{ for } i = 1, 2, \ldots, d\Big\} = \inf_{q\in\mathbb{R}^d}\Big\{P(\log \psi^q(\mathcal{A})) - \langle q, \vec{\alpha}\rangle\Big\},$$

for $\vec{\alpha} \in ri(\vec{\Omega}_{\mathcal{A}})$, *where* $ri(\vec{\Omega}_{\mathcal{A}})$ *denotes the relative interior of* $\vec{\Omega}_{\mathcal{A}}$ *(cf. [19]).*

We show that the restricted variational principle of vectors of Lyapunov exponents holds for typical cocycles.

Corollary 5.14 *Assume that* $(A_1, \ldots, A_k) \in GL_d(\mathbb{R})^k$ *generates a one-step cocycle* $\mathcal{A} : \Sigma \to GL_d(\mathbb{R})$. *Suppose that* $\mathcal{A} : \Sigma \to GL_d(\mathbb{R})$ *is a typical cocycle. Then*

$$h_{\text{top}}(E(\vec{\alpha})) = \inf_{q\in\mathbb{R}^d}\Big\{P(\log \psi^q(\mathcal{A})) - \langle \vec{\alpha}, q\rangle\Big\} = \sup\Big\{h_\mu(T) : \mu \in \mathcal{M}(\Sigma, T), \chi_i(\mu, \mathcal{A}) = \alpha_i \text{ for } i = 1, 2, \ldots, d\Big\}$$

for all $\vec{\alpha} \in ri(\vec{\Omega}_{\mathcal{A}})$.

Proof It follows from the combination of Theorem 5.12 and Theorem 5.13. □

We remark that the above corollary extend previous results about the entropy spectrum of the top Lyapunov exponent [11, 12, 17] and the entropy spectrum of certain asymptotically additive potentials [14]. Moreover, our result gives an affirmative answer to [6, Problem (7)].

Acknowledgments This text was written in connection with the conference "New Trends in Lyapunov Exponents" in Lisbon 2022. I thank the organizers and especially Pedro Duarte for the invitation to this very pleasant and stimulating week. The author was supported by the Knut and Alice Wallenberg Foundation.

References

1. L. Barreira and K. Gelfert, *Multifractal analysis for Lyapunov exponents on nonconformal repellers*, Communications in mathematical physics **267** (2006), no. 2, 393–418.
2. J. Bochi and N. Gourmelon, *Some characterizations of domination*, Mathematische Zeitschrift **263** (2009), no. 1, 221–231.
3. B. Bárány, A. Käenmäki, and I. Morris, *Domination, almost additivity, and thermodynamic formalism for planar matrix cocycles*, Israel Journal of Mathematics **239** (2020), no. 3, 173–214.

4. R. Bowen, *Markov partitions for axiom a diffeomorphisms*, American Journal of Mathematics **92** (1970), no. 3, 725–747.
5. L. Barreira and B. Saussol, *Variational principles and mixed multifractal spectra*, Transactions of the American Mathematical Society **353** (2001), no. 10, 3919–3944.
6. E. Breuillard and C. Sert, *The joint spectrum*, Journal of the London Mathematical Society **103** (2021), no. 2, 943–990.
7. C. Bonatti and M. Viana, *Lyapunov exponents with multiplicity 1 for deterministic products of matrices*, Ergodic Theory and Dynamical Systems **24** (2004), no. 5, 1295–1330.
8. Y. Cao, D. Feng, and W. Huang, *The thermodynamic formalism for sub-additive potentials*, Discrete and Continuous Dynamical Systems **20** (2008), no. 3, 639–657.
9. V. Climenhaga, *The thermodynamic approach to multifractal analysis*, Ergodic Theory and Dynamical Systems **34** (2014), no. 5, 1409–1450.
10. S. Crovisier and R. Potrie, *Introduction to partially hyperbolic dynamics*, Notes, International Centre for Theoretical Physics, Trieste, Italy (2015).
11. L. Díaz, K. Gelfert, and M. Rams, *Entropy spectrum of Lyapunov exponents for nonhyperbolic step skew-products and elliptic cocycles*, Communications in Mathematical Physics **367** (2019), no. 2, 351–416.
12. D. Feng, *Lyapunov exponents for products of matrices and multifractal analysis. part ii: General matrices.*, Israel Journal of Mathematics **170** (2009), 355–394.
13. A. Fan, D. Feng, and J. Wu, *Recurrence, dimension and entropy*, Journal of London Mathematical Society **64** (2001), no. 1, 229–244.
14. D. Feng and W. Huang, *Lyapunov spectrum of asymptotically sub-additive potentials*, Communications in Mathematical Physics **297** (2010), no. 1, 1–43.
15. B. Kalinin and V. Sadovskaya, *Cocycles with one exponent over partially hyperbolic systems*, Geometriae Dedicata **167** (2013), no. 1, 167–188.
16. R. Mohammadpour, *Entropy spectrum of Lyapunov exponents for typical cocycles*, https://arxiv.org/abs/2210.11574 (2022).
17. R. Mohammadpour, *Lyapunov spectrum properties and continuity of the lower joint spectral radius*, Journal of Statistical Physics **187** (2022), no. 3, 23.
18. R. Mohammadpour, *Restricted variational principle of Lyapunov exponents for typical cocycles*, https://arxiv.org/abs/2301.01721 (2023).
19. R. T. Rockafellar, *Convex analysis*, Princeton University Press, Princeton, N.J. (1970).

The Continuity Problem of Lyapunov Exponents

Adriana Sánchez

Abstract This is a survey of recent results on the dependence of Lyapunov exponents on the underlying data.

Keywords Random cocycles · Mixed random quasi-periodic cocycles · Lyapunov exponents · Continuity

2010 Mathematics Subject Classification Primary: 37H15; Secondary: 37A20 - 37D25

1 Introduction

The theory of Lyapunov exponents goes back to the work of Aleksadr M. Lyapunov, who studied the stability of differential equations in the nineteenth century. In Ergodic Theory, the Lyapunov exponents are quantities that measure the average exponential growth of the norm iterates of the cocycle along invariant subspaces in the fibers.

Taking as an example, the hyperbolic theory of Dynamical Systems, where one can understand certain dynamical properties of the base dynamics f by studying the action of its derivative Df on the tangent space, one can hope that by studying properties of linear cocycles one can also deduce some properties of f. Nevertheless, the notion of linear cocycle is much more general and flexible, and arises naturally in many other situations as in the spectral theory of Schrödinger operators, for instance.

The Fustenberg and Kesten theorem [22] asserts that with probability one, the logarithmic growth rate of products of random matrices equals its mean growth rate. Later, in 1968 Oseledets [28] refines Furstenberg and Kesten result providing

A. Sánchez (✉)
Centro de investigación de Matemática Pura y Aplicada, Escuela de Matemática, Universidad de Costa Rica, San José, Costa Rica
e-mail: adriana.sanchez_c@ucr.ac.cr

© The Author(s), under exclusive license to Springer Nature Switzerland AG 2023
J. L. Dias et al. (eds.), *New Trends in Lyapunov Exponents*, CIM Series in Mathematical Sciences, https://doi.org/10.1007/978-3-031-41316-2_7

a global description of all (directional) Lyapunov exponents of a linear cocycle. This two results marked the begining of the theory of Lyapunov exponents of linear cocycles.

One of the most important questions that comes up when studying Lyapunov exponents of linear cocycles is the regularity problem. By this we mean the problem of understanding the behavior of the Lyapunov exponents regarding the underlying data: measure and cocycle.

Discontinuity of Lyapunov exponents is typical for continuous cocycles taking values in $SL(2,\mathbb{R})$ over an invertible base. For instance, Bochi Mañe [8] proved that in the space of $SL(2,\mathbb{R})$-valued continuous cocycles over an aperiodic map, if a cocycle is not hyperbolic, then it can be approximated by cocycles with zero Lyapunov exponents. In particular, there are cocycles with positive Lyapunov exponents that are accumulated by cocycles with zero Lyapunov exponents. Furthermore, when the base dynamics is far from being hyperbolic, for example, when f is a rotation on the circle, Wang and You [38], showed that having non-zero Lyapunov exponents is not an open property even in the C^∞ topology.

Even though, discontinuity is a common feature, there are some contexts where continuity has been established. In this note we are going to present the most recent works developed in this direction.

The paper is organized as follows: Sect. 2 presents the principal concepts needed in order to understand the results. Section 3 is dedicated to the continuity problem for random cocycles. Finally, in Sect. 4 we introduce the most recent work regarding the continuity problem based on a new type of cocycles: the mixed random-quasi periodic cocycles.

2 Preliminaries

2.1 Linear Cocycles

Let $(M,\mathscr{B},\mu)$ be a measurable space and f a measure preserving transformation $f\colon (M,\mu)\to(M,\mu)$. Let A be a measurable function valuated in a matrix group $G\subset GL(d,\mathbb{R})$. The *linear cocycle* defined by A over f is the transformation defined by

$$F: M\times\mathbb{R}^d\to M\times\mathbb{R}^d,\quad F(x,v)=(f(x),A(x)v),$$

with iterates $F^n(x,v)=(f^n(x),A^n(x)v)$ dynamically defined for $n\geq 1$ by

$$A^n(x)=A(f^{n-1}(x))\cdots A(f(x))A(x).$$

If f is invertible, so is F and $F^{-n}(x,v)=(f^{-n}(x),A^{-n}(x)v)$ with

$$A^{-n}(x)=A(f^{-n}(x))^{-1}\cdots A(f^{-1}(x))^{-1}.$$

When the base dynamics f is fixed, we allow ourselves to call A itself a cocycle.

Example 1 Let p be a probability measure on a group $G \subset GL(d, \mathbb{R})$, $M = G^{\mathbb{N}}$, $\mu = p^{\mathbb{N}}$ and $f : M \to M$ the shift map given by

$$f((\alpha_n)_n) = (\alpha_{n+1})_n.$$

The cocycle, called the *product of random matrices*, is given by

$$A((\alpha_n)_n) = \alpha_0,$$

Note that its iterates are dynamically defined by $A^k((\alpha_n)_n) = \alpha_{k-1} \cdots \alpha_0$ for every $k \geq 1$.

Example 2 Let $f : M \to M$ be a diffeormorphism over a compact manifold M and μ be an invariant probability measure. Assuming M is parallelizable (trivial tangent bundle), the derivative cocycle of f is $A(x) = Df(x)$.

2.2 Lyapunov Exponents

The existence of the Lyapunov exponents for linear cocycles was first established by Furstenberg-Kesten Theorem [22]:

Theorem 3 (Furstenberg-Kesten) *Let $L^1(\mu)$ denote the space of μ-integrable functions on M and suppose that $\log^+ \|A^{\pm 1}\|$ belongs to $L^1(\mu)$. Then, the limits*

$$\lambda_+(A, x) = \lim_{n \to \infty} \frac{1}{n} \log \|A^n(x)\|,$$

$$\lambda_-(A, x) = \lim_{n \to \infty} \frac{1}{n} \log \|A^{-n}(x)\|^{-1},$$

exist for μ-almost every $x \in M$. We call such limits Lyapunov exponents. Moreover, $\lambda_+ \geq \lambda_-$ and are invariant functions i.e. $\lambda_\pm \circ f = \lambda_\pm$ for μ-almost every point.

A generalization of this result was given later by Kingman [26] is his ergodic theorem. This theorem also establishes that we can write

$$\lambda_+(A, x) = \inf_{n \geq 1} \frac{1}{n} \int \log \|A^n\| d\mu,$$

$$\lambda_-(A, x) = \sup_{n \geq 1} \frac{1}{n} \int \log \|(A^n)^{-1}\|^{-1} d\mu.$$

Moreover, if μ is ergodic the Lyapunov exponents are constant for μ-almost every point x and we have $\lambda_+(A, x) = \lambda_+(A, \mu)$ and $\lambda_-(A, x) = \lambda_-(A, \mu)$. Therefore, we can consider the Lyapunov exponents as functions of the measure and the cocycle.

The Multiplicative Ergodic Theorem of Oseledets [28] refines Theorem 3. It provides a global description of all directional Lyapunov exponents of a linear cocycle.

Theorem 4 (Oseledets) *Assume that* $\log^+ \|A\|$ *is integrable with respect to* μ. *Then at* μ*-almost every* $x \in M$ *there exist an integer* $k(x) \geq 1$, *a flag* $\mathbb{R}^d = V_x^1 > \cdots V_x^{k(x)} > \{0\}$, *and real numbers* $\lambda_1(x) > \cdots \lambda_{k(x)}(x)$ *such that for any* $i = 1, ..., k(x)$

- *the functions* $x \mapsto k(x), \lambda_i(x), V_x^i$ *are measurable,*
- $k(x) = k(f(x)), \lambda_i(x) = \lambda_i(f(x))$ *and* $A(x)V_x^i = V_{f(x)}^i$,
- $\lim_n \frac{1}{n} \log \|A^n(x)v\| = \lambda_i(x)$ *for every* $v \in V_x^i \backslash V_x^{i+1}$.

If the system (f, μ) *is ergodic then the functions* $x \mapsto k(x), \lambda_i(x), \dim V_x^i$*are constant* μ*-almost everywhere.*

The numbers λ_i are called the *Lyapunov exponents* of the linear cocycle, the number $m_i = \dim V^i - \dim V^{i+1}$ is called *multiplicity* of the corresponding Lyapunov exponent λ_i. The *Lyapunov spectrum* is the set of Lyapunov exponents counted with multiplicity. Furthermore, the functions λ_+ and λ_- presented in Theorem 3 coincide with the largest and the smallest Lyapunov exponents in Oseledets' theorem i.e. $\lambda_+ = \lambda_1$ and $\lambda_- = \lambda_k$. Thus we call them *extremal* Lyapunov exponents or uppper and lower Lyapunov exponents respectively.

2.3 *Continuity Problem*

As mentioned before, in this work we are concerned with the continuous dependence of the Lyapunov exponents. In order to do so we first need to define the main topology we are going to work with.

When we analyze the dependence of the Lyapunov exponents as functions of the linear cocycle we can consider two topologies: the C^0*-topology* or the *Hölder topology*. Consider the space $C^0(M)$ of continuous functions in M, endowed with the norm

$$\|A\|_0 = \sup\{\|A(x)\| : x \in M\},$$

the topology generated by this norm is called the C^0-topology.

Let $\alpha > 0$. A function $A : M \to G$ defined over a metric space M is α-Hölder if there exist a positive constant $C > 0$ such that

$$\|A(x) - A(y)\| \leq Cd(x, y)^\alpha,$$

for every $x, y \in M$. Lets denote by $H^\alpha(M)$ the set of all α-Hölder functions in M with the norm

$$\|A\|_\alpha := \sup_{x \in M} \|A(x)\| + \sup_{x \neq y} \frac{\|A(x) - A(y)\|}{d(x, y)^\alpha}.$$

On the other hand, there is more than one useful topology in the space of probability measures. For example the uniform topology or the pointwise topology, and when M is a metric space we may also consider the weak* topology (see [35]). The *weak* topology* is the smallest topology such that $\nu \mapsto \int \varphi \, d\nu$ is continuous for every bounded continuous function $\varphi : G \to \mathbb{R}$. In other words, we say that a sequence of measures $(\mu_n)_n$ converges to a measure μ in the weak* topology if and only if

$$\left| \int \psi d\mu_n - \int \psi \mu \right| \to 0,$$

when n goes to infinity, for all bounded continuous functions ψ.

One of the first results regarding the continuity problem establishes the semi-continuity of the extremal Lyapunov exponents (see [35, Lemma 9.1]) :

Lemma 5 *Consider any of the following situations:*

- *A bounded and continuous with the C^0-topology and μ with the weak* topology or,*
- *A to be bounded and measurable with the C^0-topology and μ with the pointwise topology.*

Then, the functions $(A, \mu) \mapsto \lambda_+(A, \mu)$ is upper semicontinuous and the function $(A, \mu) \mapsto \lambda_-(A, \mu)$ is lower semicontinuous.

Notice that, for any pair (A, μ), for which $\lambda_+ = \lambda_-$ at μ-almost every point, is a continuity point for the extremal Lyapunov exponents.

For further results regarding the continuity of the Lyapunov exponents, we need to make a distinction on the base dynamics. The reason is that, depending on the base different techniques are implemented. The following sections will focuses in two different types of bases: Random and mixed random quasi-periodic.

3 Random Cocycles

By a random linear cocycle we understand the skew-product dynamical system defined by a Bernoulli or a Markov shift on the base and a locally constant linear fiber map.

Let p be a probability measure on a group $G \subset GL(d, \mathbb{R})$, $M = G^{\mathbb{N}}$, $\mu = p^{\mathbb{N}}$ and $f : M \to M$ the shift map as in example 1. The map f determines an ergodic transformation on (M, μ) called the *Bernoulli shift*. If we consider the space of bi-sided sequences $M = G^{\mathbb{Z}}$ and the measure $\mu = p^{\mathbb{Z}}$ it is called the *full Bernoulli shift*.

The Bernoulli shift basically models a sequence of identical experiments where the result of every experiment is also independent of the others. We can generalize the class of Bernoulli shifts by dropping off the independence condition, by assuming a dependence on the immediately previous result.

Assume we have what is called a family of *transition probabilities*. That is, we have $\{P(x, \cdot) : x \in G\}$ a family of probabilities in G depending measurably on the point x. So, given a measurable set E the number $P(x, E)$ represents the probability of the point $x_{n+1} \in E$ knowing that $x_n = x$. Assume there exist a probability measure p in G such that

$$\int P(x, E)dp(x) = p(E), \text{ for every measurable set } E \subset G. \tag{1}$$

Fixing p, we can define a measure $\tilde{\mu}$ such that, for every cylinder $[m : A_m, \ldots, A_n]$ of M, it satisfies

$$\begin{aligned}
&\tilde{\mu}\left([m : A_m, \ldots, A_n]\right) \\
&\quad = \int_{A_m} dp(x_m) \int_{A_{m+1}} dP(x_m, x_{m+1}) \cdots \int_{A_n} dP(x_{n-1}, x_n).
\end{aligned}$$

This is an invariant measure for f and the system $(f, \tilde{\mu})$ is called *Markov shift*. Notice that, when $P(x, \cdot)$ does not depend on the point x we have again the Bernoulli shift.

A *random cocycle* is a linear cocycle $A : M \to GL(d, \mathbb{R})$ over a Bernoulli or a Markov shift f given by

$$A((x_n)_n) = A_{x_0}, \tag{2}$$

where $A_1, \ldots, A_s \in GL(d, \mathbb{R})$. These are also called *locally constant cocycles*. For the rest of this section we fix the cocycle as given by (2) and consider the Lyapunov exponents as functions of the measure.

3.1 Continuity

One of the first milestones of this theory was Furstenberg's formula in [21], which expresses the top Lyapunov exponent λ_+ via an integral formula involving a so-called stationary measure of the random linear cocycle. Given a probability distribution ν on G, a probability measure η on the projective space in $\mathbb{P}\mathbb{R}^d$ is called

ν-stationary if

$$\eta(E) = \int_G \eta\big(g^{-1}(E)\big)\, d\nu(g) \text{ for every measurable set } E \subset \mathbb{P}\mathbb{R}^d. \tag{3}$$

One can check that η is ν-stationary if and only if the product measure $\nu^{\mathbb{N}} \times \eta$ is invariant under the projective cocycle. That is, the cocycle generated by the projective class of A:

$$\mathbb{P}F : G^{\mathbb{N}} \times \mathbb{P}\mathbb{R}^d \to G^{\mathbb{N}} \times \mathbb{P}\mathbb{R}^d, \quad \mathbb{P}F\big((g_n)_n, \xi\big) = \big((g_{n+1})_n, g_0\xi\big).$$

The set of stationary measures $\mathrm{Stat}(\nu)$ is a non-empty, convex, compact subset of the space of probability measures on $\mathbb{P}\mathbb{R}^d$, equipped with the weak* topology. Furthermore, the map $\nu \mapsto \mathrm{Stat}(\nu)$ is upper semicontinuous with respect to the Hausdorff topology on the space of compact subsets of $\mathscr{M}(\mathbb{P}\mathbb{R}^d)$. In other words, if $\eta_k \in \mathrm{Stat}(\nu_k)$ for every k and the sequence $(\nu_k)_k$ converges to ν then every accumulation point of $(\eta_k)_k$ is contained in $\mathrm{Stat}(\nu)$.

The Furstenberg's formula (see [35, Proposition 6.7]) establishes that

$$\lambda_+(\nu) = \max\left\{ \int_{G\times\mathbb{P}\mathbb{R}^d} \phi\, d(\nu \times \eta) : \eta \in \mathrm{Stat}(\nu)\right\}, \tag{4}$$

where $\phi : G \times \mathbb{P}\mathbb{R}^d \to \mathbb{R}$ is the *dilation function* defined by

$$\phi(g, v) = \log \frac{\|g v\|}{\|v\|}, \tag{5}$$

where v represents both a non-zero vector (on the right) and the corresponding element of the projective space $\mathbb{P}\mathbb{R}^d$ (on the left).

In [23] H. Furstenberg and Y. Kifer used this formula to prove the continuity of the (top) Lyapunov exponent under a generic irreducibility assumption on the random linear cocycle. We call ν *irreducible* if the matrices $g \in \operatorname{supp}\nu$ have no common invariant subspace and, *strongly irreducible* if there is no finite family of subspaces of $\mathbb{R}^d$ invariant under every g in the support of ν. They proved that irreducibility implies that there exist only one stationary measure. That is, if we have a sequence $(\nu_k)_k$ of probability measures on G converging to an irreducible measure ν in the weak* topology, if $(\eta_k)_k$ is a sequence of ν_k-stationary measures achieving the maximum in (4), up to restricting to a subsequence, η_k converges to some ν-invariant measure η which is unique. Then, by Furstenberg formula

$$\lambda_+(\nu_k) = \int \phi d\,(\nu_k \times \eta_k) \to \int \phi d\,(\nu \times \eta) = \lambda_+(\nu). \tag{6}$$

Therefore, it follows that the largest Lyapunov exponent is continuous at every irreducible measure. We recommend [35, Chapter 6] for a proof assuming strong irreducibility. In [23] the conclusion is extended to the case when ν is only quasi-irreducible, meaning that there exists at most one invariant subspace.

When we deal with measures that are not irreducible the situation becomes a whole lot more subtle. An example given by Kifer [25] clarifies this.

Example 6 Consider the probability measure $\mu = p_1\delta_{\alpha_1} + p_2\delta_{\alpha_2}$, with $p_1, p_2 > 0$ such that $p_1 + p_2 = 1$ and

$$\alpha_1 = \begin{pmatrix} 2^{-1} & 0 \\ 0 & 2 \end{pmatrix}, \quad \text{and} \quad \alpha_2 = \begin{pmatrix} 0 & -1 \\ 1 & 0 \end{pmatrix}.$$

Then is clear that $\lambda_+ = 0$ for every positive weight p_2. However, if p_2 tends to 0 then λ_+ tends to $\log 2$.

Regarding the problem of general continuity of the Lyapunov exponents, Bocker-Neto and Viana [9] proved that this continuity always holds for random $GL(2, \mathbb{R})$-valued linear cocycles at least for probability measures with compact support in $GL(2, \mathbb{R})$. To avoid the problem presented in the Kifer example above, they introduce a new topology where two compactly supported probability measures are close if and only if they are weak*-close and their supports are Hausdorff-close.

Theorem 7 (Bocker, Viana) *Let $\mathscr{G}(2)$ be the space of probability measures with compact support in $GL(2, \mathbb{R})$, with the smallest topology, $\mathscr{T}$, such that*

- *$p \mapsto \int \varphi dp$ is continuous for every continuous function $\varphi : G \to \mathbb{R}$ and*
- *$p \mapsto \operatorname{supp} p$ is continuous relatively to the Hausdorff topology.*

Then, the Lyapunov exponents $\mu \mapsto \lambda_\pm(\mu)$ vary continuosly in $\mathscr{G}(2)$.

The idea of their proof is to assume we have a discontinuity point ν of λ_+. So, if $(\nu_k)_k$ is a sequence converging to ν and $(\eta_k)_k$ is a sequence of ν_k-stationary measures achieving the maximum in (4), then the limit presented in (6) is strictly smaller than $\lambda_+(\nu)$. Then they show that in these circumstances η can not be the limit of stationary measures η_k for probability measures ν_k converging to ν. In order to do so they use the deterministic filtration theorem in [23] to construct an atom to η which is also a local *repeller* E (known as the equator). That kind of repelling behavior persists for nearby probability distributions ν_k, so that the corresponding stationary measures η_k should not be able to accumulate too much mass on the vicinity of E. Then an atom at E could not exist for any weak*-limit measure η (see [35, Chapter 10]).

A similar result was recently proven by Avila et al. [1] to hold for random $GL(d, \mathbb{R})$-valued cocycles for any $d \geq 2$.

Theorem 8 (Avila, Eskin, Viana) *The map $\nu \mapsto (\lambda_1, ..., \lambda_d)$ is continuous on $\mathscr{G}(d)$, for every $d \geq 2$.*

The proof of Theorem 8 follows the main idea of Bocker Viana's proof, although the approach they use is slightly different. This method is called *energy argument*, we refer the reader to [36] for a proof in the case $d = 2$.

The assumption of compact support in Theorem 7 is fundamental. As proved by Viana Sánchez [33], without the compactness assumption the weak* topology is not enough to even guarantee semi-continuity. Moreover, they proved that λ_+ and λ_- are not upper and lower semicontinuous respectively relative to the weak* topology. However, they where able to regain this property using a stronger topology.

The **Wasserstein space** is the space of probability measures which have a finite moment of order 1. By this we mean the space

$$P_1(M) := \left\{ \mu \in P(M) : \int_M d(x_0, x) d\mu(x) < +\infty \right\},$$

where $x_0 \in M$ is arbitrary and $P(M)$ denotes the space of Borel probability measures on M. This does not depend on the choice of the point x_0. Notice that the Lyapunov exponents for measures in $P_1(M)$ always exist since $\log \|A\|$ is μ-integrable for every $\mu \in P_1(M)$.

We say that a sequence μ_k converges in the *Wasserstein topology* to μ, if

- $\mu_k \xrightarrow{*} \mu$,
- $\int d(x_0, x) d\mu_k(x) \to \int d(x_0, x) d\mu(x)$.

This convergence is metrizable by the Wasserstein distance

$$W_1(\mu, \nu) = \sup \left\{ \int_M \psi d\mu - \int_M \psi d\nu \right\}, \tag{7}$$

where the supremum in the right hand side is taken over the 1-Lipschitz functions ψ.

With this topology Sánchez Viana proved in [33] the following

Theorem 9 (Sánchez Viana) *The function $\lambda_+ : P_1(SL(2, \mathbb{R})) \to \mathbb{R}$ is upper semicontinuous with the Wasserstein topology. The same is true for the function λ_- with lower semicontinuity.*

The idea of their proof takes advantage of both the Furstenberg's formula (4) and the properties of Wasserstein topology. The Furstenberg's formula allows them to consider the integrals of the dilatation function (5), while using the properties of the Wasserstein topology allows them to control the integrals outside a compact set. However, the Wasserstein topology is not strong enough to guarantee the continuity of the Lyapunov exponents for measures with no compact support, so it remains an open question.

In the same direction of Theorem 7, Malheiro and Viana [27] proved the continuity of the Laypunov exponents for random $GL(2, \mathbb{R})$-valued linear cocycles over mixing Markov shifts by using similar arguments. They consider a finite set $G = \{A_1, ...A_s\}$ of matrices in $GL(2, \mathbb{R})$, with a probability vector $p = (p_1, ..., p_s)$ and transition probabilities P_{ij} with $i, j = 1, ..., s$. The key ingredient in their proof is the definition of stationary measures in their context. A vector $\eta = (\eta_i)_i$ of probability measures in $\mathbb{PR}^2$ is said P-stationary if

$$\sum_{i \in X} p_i P_{ij} \eta_i \left(A(i)^{-1}(D) \right) = p_j \eta_j(D)$$

for every $j = 1, ..., s$ and every measurable set $D \subset \mathbb{PR}^2$. After proving that the set $\{(P, \eta) : \eta$ is a P-stationary vector of probabilities$\}$ is closed, they proceed as in the proof of Theorem 7 by constructing an atom that is also a local repeller for η and extending Bocker Viana's estimatives for the vector case.

These results were extended beyond the class of random cocycles by Bakes et al. [3]. They proved that the continuity of Lyapunov exponents holds when restricted to the realm of fiber-bunched Hölder cocycles over any hyperbolic system and for any ergodic probability measure with local product structure. More recently, Viana and Yang [37] were able to prove the continuity of Lyapunov exponents in the C^0 topology for a subset of linear cocycles when the transformation in the base is a uniformly expanding map. In [20] Freijo Marin proved the continuity of the Lyapunov exponents for non-uniformly fiber-bunched cocycles in $SL(2, \mathbb{R})$.

3.2 *Hölder Continuity*

There are several works regarding the problem of regularity of Lyapunov exponents. In the case of the Bernoulli shift, Le Page [29] proved that the top Lyapunov exponent is actually locally Hölder continuous if we also assume a contraction condition. By locally Hölder continuous function we mean a function which is Hölder in a neighborhood of each point of its domain.

Let $\sigma_1(g) \geq ... \geq \sigma_d(g)$ denote the singular values of a matrix $g \in GL(d, \mathbb{R})$. That is, they are the roots of the eigenvalues of the positive define self-adjoint matrix g^*g. We say that a measure ν in G has the *contraction property* if the quotient $\sigma_1(g)\backslash\sigma_2(g)$ is unbounded on the closed semi-group generated by the support of ν.

Theorem 10 (Le Page) *If a compactly supported measure ν in $GL(d, \mathbb{R})$ is strongly irreducible and has the contraction property then the upper Lyapunov exponent is locally Hölder continuous.*

His proof is based on a spectral method, which exploits the existence of a gap in the spectrum of a certain Markov operator that traces the action of the linear cocycle on the projective space. The existence of a spectral gap is forced by the generic hypothesis of Le Page's continuity theorem. Baraviera Duarte [4] gave a new proof of Le Page's Theorem (see also [16]). Their approach suggests an algorithm to approximate the upper Lyapunov exponent and the stationary measure for such random cocycles.

The Hölder modulus of continuity in Le Page's theorem 10 is optimal as proven by an example due to Halperin (see [32, Appendix 3]). This example consist of the

1-parameter family of measures supported in $SL(2, \mathbb{R})$ defined as follows

$$\mu_{a,b,E} := \frac{1}{2}\left(\delta_{A_E} + \delta_{B_E}\right)$$

where A_E and B_E are the matrices

$$A_E = \begin{pmatrix} a - E & -1 \\ 1 & 0 \end{pmatrix}, \quad \text{and} \quad B_E = \begin{pmatrix} b - E & -1 \\ 1 & 0 \end{pmatrix},$$

for some parameters $a, b \in \mathbb{R}$. It follows from [32, Appendix 3]) that the upper Lyapunov exponent $E \mapsto \lambda_+(\mu_{a,b,E})$ is not Hölder continuous of any order α larger than

$$\alpha_0 = \frac{\log 2}{arc \cosh\left(1 + \frac{1}{2}|a - b|\right)}.$$

Furthermore, the measures $\mu_{a,b,E}$ satisfies the assumptions in Le Page's theorem. This implies that $E \mapsto \lambda_+(\mu_{a,b,E})$ are necessarily Hölder continuous for small α.

Duarte and Klein [15] removed the irreducibility hypotesis in Le Page's theorem for finitely supported measures in $GL(2, \mathbb{R})$. They proved that if a finitely supported measure has distinct Lyapunov exponents, the function $\mu \mapsto \lambda_+(\mu)$ is either locally Hölder or else locally weak-Hölder.

For the time being, consider probability measures finitely supported. That is, measures of the form

$$\nu = p_1\delta_{g_1} + \cdots + p_N\delta_{g_N}, \tag{8}$$

with $p_1, ..., p_N > 0$ and $g_1, ..., g_n \in GL(2, \mathbb{R})$. In this case we say μ is a *finitely supported measure* in $GL(2, \mathbb{R})$ since its support is in $GL(2, \mathbb{R})$.

Take the distance between two such measures as to be defined by

$$d(\nu, \nu') = \max\{|p_i - p_i'|, \|g_i - g_i'\| : i = 1, ..., N\}.$$

Theorem 11 (Duarte Klein) *Let ν be a finitely supported measure in $GL(2, \mathbb{R})$ over the full Bernoulli shift. That is the shift defined in the space $G^{\mathbb{Z}}$. If the measure ν is such that $\lambda_-(\nu) < \lambda_+(\nu)$. Then, if the support of ν is diagonalizable, there exist constants $C > 0$, $\alpha > 0$ and $\beta \in (0, 1]$ such that*

$$d(\lambda_+(\nu) - \lambda_+(\nu')) \le C \exp\left(-\alpha \left(\log \frac{1}{d(\nu, \nu')}\right)^{\beta}\right),$$

for every ν' in a neighborhood of ν. If the support of ν is not diagonalizable, λ_+ is Hölder continuous.

Moreover, in a recent work of Duarte et al. [18] the authors proved that in the reducible case the modulus of continuity of the Lyapunov exponents deteriorate drastically. They construct an example of a Bernoulli shift near the Kifer Example 6 such that its Lyapunov exponent cannot be Hölder or weak-Hölder continuous, thus providing a limitation on the modulus of continuity of the Lyapunov exponent of random cocycles. They consider the measure $\mu := \frac{1}{2}(\delta_A + \delta_B)$ where

$$A = \begin{pmatrix} 0 & -1 \\ 1 & 0 \end{pmatrix}, \quad \text{and} \quad B = \begin{pmatrix} e & 0 \\ 0 & e^{-1} \end{pmatrix},$$

and proved that there exist a curve $\tilde{\mu}_t = \frac{1}{2}\left(\delta_{A_t} + \delta_{B_t}\right)$ with $\tilde{\mu}_0 = \mu$ such that $t \mapsto \lambda_+(\tilde{\mu}_t)$ is not weak-Hölder around $t = 0$. In this case, μ does not satisfy Le Page's assumptions.

In a recent work of Bezerra Duarte [6] they proved that if μ is a finitely supported measure on $SL(2, \mathbb{R})$ with positive Lyapunov exponent but not uniformly hyperbolic, then the Lyapunov exponent function is not α-Hölder around μ for any α exceeding the Shannon entropy of μ over the Lyapunov exponent of μ. Here *uniformly hyperbolic cocycle* means that there exist constants $C > 0$ and $\gamma > 0$ such that for every $n \geq 1$ and $x \in M = G^{\mathbb{Z}}$ then

$$\|A^n(x)\| \geq Ce^{\gamma n}.$$

Theorem 12 (Bezerra Duarte) *Let μ be a finitely supported measure on $SL(2, \mathbb{R})$. Assume that $\lambda_+(\mu) > 0$, μ is irreducible and has heteroclinic tangency. Then, there exists an analytic one parameter family of finitely supported measures $(\mu_t)_t$ such that $\mu_0 = \mu$ and for any $\alpha > \frac{H(\mu)}{\lambda_+(\mu)}$, the function $t \mapsto \lambda_+(\mu_t)$ is not locally α-Hölder continuous at any neighborhood of $t = 0$.*

Here $H(\mu)$ denotes the *Shannon's entropy*

$$H(\mu) = -\sum_{i=1}^{N} p_i \log p_i.$$

We say that μ has a heteroclinic tangency if there are matrices A, B and C in the semigroup generated by the support of μ such that A and B are hyperbolic and $C\hat{u}(B) = \hat{s}(A)$. Where $\hat{u}(B)$ represents the unstable direction of the hyperbolic matrix B and, $\hat{s}(A)$ the stable direction of A.

In their proof they follow the same stategy presented in Halperin's example [32] by embedding the cocycle A into a family of locally constant Schrödinger cocycles over a Markov shift. Then, they use the heteroclinic tangency to map the horizontal direction into the vertical direction after several iterations and break the Hölder regularity.

For the general case of measures with compact support (not necessary finite), consider the topology $\mathscr{T}$ defined in Theorem 7. In the compact setting, the Wasserstein topology coincides with the weak* topology, so the Wasserstein distance presented in (7) jointly with the Haussdorff distance for compact spaces

$$d_H(K_1, K_2) = \inf\{r > 0 : K_1 \subset B_r(K_2), \text{ and } K_2 \subset_r (K_1)\}, \tag{9}$$

metricize the topology $\mathscr{T}$. That is,

$$d_{\mathscr{T}}(\nu, \nu') = W_1(\nu, \nu') + d_H(\operatorname{supp}\nu, \operatorname{supp}\nu')$$

is a metric in $\mathscr{G}(2)$ compatible with the topology $\mathscr{T}$. It is worth to mention that in the Hausdorff distance (9) the neighborhoods B_r are taken with respect to the distance on G given by

$$d_G(g_1, g_2) = |g_1 - g_2| + \left|g_1^{-1} - g_2^{-1}\right|.$$

Using this topology, Tall Viana [34] proved that for random $GL(2, \mathbb{R})$ cocycles the Lyapunov exponents are point-wisely Hölder when the Lyapunov exponents are distinct.

Theorem 13 (Tall Viana) *For every $\nu \in \mathscr{G}(2)$ with $\lambda_-(\nu) < \lambda_+(\nu)$ there exist constants $C > 0$ and $\beta > 0$ and a neighborhood $U \subset \mathscr{G}(2)$ such that*

$$\left|\lambda_\pm(\nu) - \lambda_\pm(\nu')\right| \le C d_{\mathbb{T}}(\nu, \nu')^\beta \text{ for every } \nu' \in U.$$

Their proof focuses in proving the existence of a unique maximal ν-stationary measure which is some kind of hyperbolic attractor for the random walk defined by ν Moreover, they also proved that the Lyapunov exponents are *log-Hölder* continuous at every point of $\mathscr{G}(2)$.

Theorem 14 (Tall Viana) *For every $\nu \in \mathscr{G}(2)$ there exist constants $C > 0$ and $\beta > 0$ and a neighborhood $U \subset \mathscr{G}(2)$ such that*

$$\left|\lambda_\pm(\nu) - \lambda_\pm(\nu')\right| \le C\left(\log \frac{1}{d_{\mathbb{T}}(\nu, \nu')}\right)^{-\beta}, \text{ for every } \nu' \in U.$$

For this result they proved that when $\lambda_+(\nu) = \lambda_-(\nu)$ the random walk has power law diffusion to infinity (see [36]).

This result was extended to the linear cocycles over a hyperbolic base by Duarte et al. [17]. They proved that locally near any typical cocycle, the Lyapunov exponents are Hölder continuous functions relative to the uniform topology.

Recently, Duarte Freijo [19] generalized the previous results for non-invertible random cocycles. They established uniform large deviations estimates of exponential type and Hölder continuity of the Lyapunov exponents for random non-invertible cocycles with constant rank.

3.3 Analyticity

A classical theorem of Ruelle [31, Theorem 3.1] proves the analyticity of the Lyapunov exponent for uniformly hyperbolic $SL(2, \mathbb{R})$ cocycles and, more generally, for $GL(d, \mathbb{R})$ cocycles with dominated splitting and 1-dimensional strongly unstable direction.

Later, Peres [30] proved that for random cocycles generated by locally constant $GL(d, \mathbb{R})$-valued functions with finitely many values, if the upper Lyapunov exponent is simple, it is locally analytic as a function of the probability weights.

Theorem 15 (Peres) *Let μ be a finitely supported measure in $GL(d, \mathbb{R})$ (as in (8)). Assume that the k-th Lyapunov exponent is simple i.e. $\lambda_{k+1} > \lambda_k > \lambda_{k-1}$ and the probability weights in (8) satisfy that $p_i > 0$ for $i = 1, .., N$. Then, λ_k is a locally real analytic function of $p = (p_1, ..., p_N)$.*

The idea of the proof of Peres' Theorem 15 is to combine some stability results due to Furstenberg and Kifer [23] and the analysis of Le Page [29]. In fact, he takes advantage of Furstenberg formula 4 and, the hyperbolic estimates established by Le Page to reduce the proof to the irreducible case by a recurrence argument.

4 Mixed Random-Quasiperiodic Cocycles

An important type of cocycles presented in mathematical physics are the *quasi-periodic cocycles*. A quasi-periodic cocycle is a linear cocycle over a fixed ergodic torus translation of one or several variables, where the fiber action depends analytically on the base point. That is, a skew product map of the form

$$\mathbb{T}^m \times \mathbb{R}^d \ni (t, v) \mapsto (\tau_\alpha(t), A(t)v) \in \mathbb{T}^m \times \mathbb{R}^d,$$

where $\tau_\alpha(t) = t + \alpha$ is a translation on $\mathbb{T}^m$ by rationally independent frequency $\alpha \in \mathbb{T}^m$ and $A : \mathbb{T}^m \to SL(d, \mathbb{R})$ a continuous matrix valued function on the torus.

Thus, the quasi-periodic cocycle can be identified with the pair (α, A). Let us denote by

$$\mathscr{G} = \mathscr{G}(m, d) := \mathbb{T}^m \times C^0(\mathbb{T}^m, SL(d, \mathbb{R})) \tag{10}$$

be the set of all quasi-periodic cocycles.

It is well-known that good regularity behavior of the Lyapunov exponent can give information about the properties of the associated Schrödinger operator (see for example [12, 13]). However, it is extremely hard to obtain regularity properties of the Lyapunov exponents in this context as one can see from [2, 14]. This led to the introduction of a mixed model of random product of quasi-periodic cocycles called the mixed random quasi-periodic cocycles, that allow us to transfer good regularity properties that comes from the random realm to the quasi-periodic world.

Let $(\Omega, \mathscr{B})$ be a Polish space (a separable, completely metrizable topological space) and $\mathscr{B}$ is its Borel σ-algebra. Let ν be a compactly supported Borel probability measure on Ω. Consider the Bernoulli shift on $M = \Omega^{\mathbb{N}}$ (or the full Bernoulli shift on $M = \Omega^{\mathbb{Z}}$). Let $\mathbb{T}^m$ be the torus of dimension m, and denote by $\hat{m}$ the Haar measure on its Borel σ-algebra. Given a continuous $\mathfrak{a} : \Omega \to \mathbb{T}^m$, the skew product map

$$g : M \times \mathbb{T}^m \to M \times \mathbb{T}^m, \quad g(x,t) := (f(x), t + \mathfrak{a}(x_0)) \tag{11}$$

will be referred to as a *mixed random quasi-periodic* base dynamics. This maps preserves the measure $\nu^{\mathbb{N}} \times \hat{m}$.

Example 16 Given a frequency $\alpha \in \mathbb{T}^m$, let $\mathfrak{a} : \Omega \to \mathbb{T}^m$ be the constant function $\mathfrak{a}(x_0) \equiv \alpha$. Then the corresponding skew product map g on $\Omega^{\mathbb{N}} \times \mathbb{T}^m$ is given by

$$g(x,t) = (f(x), t + \alpha).$$

Thus, it is the product between a Bernolli shift and the torus translation by α.

Consider a compactly supported probability measure ν on $\mathscr{G}$. Let $\Omega \subset \mathscr{G}$ be a closed subset such that $\operatorname{supp} \nu \subset \Omega$. Consider the linear cocycle $F : M \times \mathbb{T}^m \times \mathbb{R}^d \to M \times \mathbb{T}^m \times \mathbb{R}^d$ defined by

$$F(x,t,v) := (f(x), t + \mathfrak{a}(x_0), \mathscr{A}(x_0)(t)v)\,,$$

where $\mathscr{A} : \Omega \to C^0(\mathbb{T}^m, SL(d,\mathbb{R}))$. The invertible version of this map, with the same expression, is defined on $\Omega^{\mathbb{Z}} \times \mathbb{T}^m \times \mathbb{R}^d$.

Consequently, the base dynamics of the cocycle F is the mixed random quasi-periodic map presented in (11), while the fiber action is induced by the map $\mathscr{A}(x,t) =: \mathscr{A}(x_0)(t) \in SL(d,\mathbb{R})$. The skew product F will then be referred to as a *mixed random quasi-periodic cocycle*.

Consider the case when $\mathfrak{a}(x_0) = \theta_{x_0} \in (0,1]$ irrational. In particular, the base dynamics with the full shift and the measure $\nu^{\mathbb{Z}} \times \mathrm{Leb}$ is an ergodic system. Bezerra Poletti [5] proved that the set of random product of $k+1$ C^r, $0 \le r \le \infty$ quasi-periodic cocycles for which the associated largest Lyapunov exponent of the random product is positive, contains a C^0 open and C^r dense subset which is formed by C^0 continuity point of the Lyapunov exponent.

Theorem 17 (Bezerra Poletti) *For $r \in [0,\infty] \cup \{\omega\}$ there exist a C^0 open and C^r dense subset of $(C^r(\mathbb{T}, SL(2,\mathbb{R})))^{k+1}$, such that the random product defined by cocycles in this set has positive Lyapunov exponent and is a C^0-continuity point for the Lyapunov exponents.*

To prove they result they realize that a special class of random quasi-periodic cocycles is C^r dense. That special class is the class of weakly simple cocycles. That is, the set of cocycles $\mathscr{A}$ such that they are weakly pinching and weakly twisting. For the precise definition of these terms we refer the reader to [5].

Actually, the first time the term mixed random quasi-periodic cocycle was coined was in the work of Cai et al. [10]. They characterized the ergodicity of the base dynamics and establish a large deviations type estimate for certain types of observables. Moreover, for the fiber dynamics they proved the uniform upper semicontinuity of the maximal Lyapunov exponent as a function of the measure, relative to the Wasserstein distance. Fix a number $L < \infty$, let $\Omega = \mathscr{G}_L$ the set of quasi-periodic cocycles (α, A) such that $\|A\|_0 < L$.

Theorem 18 (Cai, Duarte, Klein) *Let ν_0 a compactly supported measure on Ω be an ergodic measure with respect to g, a mixed random quasi-periodic base dynamics. Given any $\varepsilon > 0$ there are $\delta = \delta(\varepsilon, \nu_0, L) > 0$, $\bar{n} = \bar{n}(\varepsilon, \nu_0, L) \in \mathbb{N}$ and $c = c(\varepsilon, \nu_0, L) > 0$ such that for all compactly supported measure ν on Ω with $W_1(\nu, \nu_0) < \delta$, for all $t \in \mathbb{T}^m$ and for all $n \geq \bar{n}$ we have*

$$\nu^{\mathbb{Z}} \left\{ x \in M : \frac{1}{n} \log \left| \mathscr{A}^n(x)(t) \right| \geq \lambda_+(\nu_0) + \varepsilon \right\} < e^{-cn}.$$

Moreover, the map $\nu \mapsto \lambda_+(\nu)$ is upper semicontinuous with respect to the Wasserstein metric in the space of ergodic measures.

Furthermore, Caio, Duarte, Klein in [11] extended the results of Furstenberg's theory on random products of matrices to this more general setting. For example, they derived a Furstenberg-type formula to represent the maximal Lyapunov exponent and established its continuity under some generic assumptions. These results were already proven in a setting that includes this one by Kifer [24]. For example, Kifer proved the following Furstenberg-type formula to represent the maximal Lyapunov exponent:

Proposition 19 (Kifer) *Let $\mathscr{V}$ be an (A, p)-invariant measurable section then*

$$\lambda_+(A, p) = \max\{\lambda_+(A^{\mathscr{V}}, p), \lambda_+(A^{\mathbb{R}^d/\mathscr{V}}, p)\}.$$

Recently, Bezerra et al. [7] extended Peres conclusions in [30] to include this class of cocycles. That is, they proved the following.

Theorem 20 (Bezerra, Sánchez, Tall) *Let $p = (p_1, \ldots, p_N) \in \mathbb{R}^N$ be a probability vector such that $p_i > 0$ for every $i = 1, \ldots, N$. If $\lambda_+(p)$ is simple, then the function which associates each probability vector q the (top) Lyapunov exponent $\lambda_+(q)$ can be extended to an analytic function in a neighborhood of p.*

In their proof they use the concept of invariant sections. A measurable section $\mathscr{V} : \mathbb{T}^m \to Gr(k, d)$, where $Gr(k, d)$ denotes the Grassmanian of the k-dimensional subspaces inside of $\mathbb{R}^d$, is a *(A, p)-invariant section* if

$$A_x(t)\mathscr{V}(t) = \mathscr{V}(f_x(t)),$$

for μ_p-a.e. $(x, t) \in \Sigma \times \mathbb{T}^m$. In other terms, we have that

$$A_j(t)\mathscr{V}(t) = \mathscr{V}(t + \theta_j)$$

for $j = 1, \ldots, N$, and Leb-a.e. $t \in \mathbb{T}^m$.

Given an invariant section $\mathscr{V}$ one can define a canonical linear cocycle associated to $\mathscr{V}$. Let $J_{\mathscr{V}}(t)$ be the appropriate change of coordinates. We can define a linear cocycle $F_{\mathscr{V}}$ in $M \times \mathbb{T}^m \times \mathbb{R}^k$ by

$$F_{\mathscr{V}}(x, t, v) = (\sigma(x), f_{x_0}(t), A^{\mathscr{V}}_{x_0}(t)v),$$

where $A^{\mathscr{V}}_i : \mathbb{T}^m \to GL(2, \mathbb{R})_k(\mathbb{R})$ is given by

$$A^{\mathscr{V}}_i(t) := J_{\mathscr{V}}(t + \theta_i) \circ A_{x_0}(t) \circ (J_{\mathscr{V}}(t))^{-1},$$

for every $i = 1, \ldots, N$ and $A^{\mathscr{V}} = (A^{\mathscr{V}}_1, \ldots, A^{\mathscr{V}}_N)$.

Moreover, we can consider the factor space $\mathbb{T}^m \times \mathbb{R}^d/\mathscr{V}$ where each two points (t, u) and $(t, v) \in \mathbb{T}^m \times \mathbb{R}^d$ are identified if $u - v \in \mathscr{V}(t)$. Notice that $A_x(t)$ acts on $\mathbb{R}^d/\mathscr{V}(t)$ for any t, and since $\mathscr{V}$ is (A, p)-invariant we have that

$$A_x(t)(\mathbb{R}^d/\mathscr{V}(t)) = \mathbb{R}^d/\mathscr{V}(f_x(t)),$$

for μ_p-a.e. $(x, t) \in \Sigma \times \mathbb{T}^m$. As before, we can define a linear cocycle $F_{\mathbb{R}^d/\mathscr{V}}$ in $M \times \mathbb{T}^m \times \mathbb{R}^{d-k}$ by

$$F_{\mathbb{R}^d/\mathscr{V}}(x, t, v) = (\sigma(x), f_{x_0}(t), A^{\mathbb{R}^d/\mathscr{V}}_{x_0}(t)v),$$

where $A^{\mathbb{R}^d/\mathscr{V}}_i : \mathbb{T}^m \to \mathrm{GL}(d - k, \mathbb{R})$ is defined using the appropriate change of coordinates and $A^{\mathbb{R}^d/\mathscr{V}} = (A^{\mathbb{R}^d/\mathscr{V}}_1, \ldots, A^{\mathbb{R}^d/\mathscr{V}}_N)$.

Using the Furstenberg-type formula in Proposition 19 and, following Peres [30] proof, they reduce the proof of their result to the irreducible case which they conclude by considering the hyperbolic estimates presented in [5].

Acknowledgments The author was supported by Universidad de Costa Rica.

References

1. A. Avila, A. Eskin, and M. Viana. Continuity of the Lyapunov exponents of random matrix products. *arXiv preprint arXiv:2305.06009*, 2023.
2. Artur Avila, Svetlana Jitomirskaya, and Christian Sadel. Complex one-frequency cocycles. *Journal of the European Mathematical Society*, 16(9):1915–1935, 2014.

3. Lucas Henrique Backes, Aaron Brown, and Clark Butler. Continuity of Lyapunov exponents for cocycles with invariant holonomies. *Journal of Modern Dynamics. JMD. Singapura: AIMS, 2018. Vol. 12,(June 2018), p. 261–283*, 2018.
4. Alexandre Baraviera and Pedro Duarte. Approximating Lyapunov exponents and stationary measures. *Journal of Dynamics and Differential Equations*, 31(1):25–48, 2019.
5. J. Bezerra and M. Poletti. Random product of quasi-periodic cocycles.
6. Jamerson Bezerra and Pedro Duarte. Upper bound on the regularity of the Lyapunov exponent for random products of matrices. *arXiv preprint arXiv:2208.03575*, 2022.
7. Jamerson Bezerra, Adriana Sánchez, and El Hadji Yaya Tall. Analyticity of the Lyapunov exponents of random products of quasi-periodic cocycles. *arXiv preprint arXiv:2111.00683*, 2021.
8. Jairo Bochi. Genericity of zero Lyapunov exponents. *Ergodic Theory and Dynamical Systems*, 22(6):1667–1696, 2002.
9. C. Bocker and M. Viana. Continuity of Lyapunov exponents for random two-dimensional matrices. *Ergodic Theory Dynam. Systems*, 37:1413–1442, 2017.
10. Ao Cai, Pedro Duarte, and Silvius Klein. Mixed random-quasiperiodic cocycles. *arXiv preprint arXiv:2109.09544*, 2021.
11. Ao Cai, Pedro Duarte, and Silvius Klein. Furstenberg theory of mixed random-quasiperiodic cocycles. *arXiv preprint arXiv:2201.04745*, 2022.
12. D. Damanik. Schrödinger operators with dynamically defined potentials. *Ergodic Theory Dynam. Systems*, 37:1681–1764, 2017.
13. David Damanik. Lyapunov exponents and spectral analysis of ergodic Schrödinger operators: a survey of Kotani theory and its applications. In *Spectral theory and mathematical physics: a Festschrift in honor of Barry Simon's 60th birthday*, volume 76 of *Proc. Sympos. Pure Math.*, pages 539–563. Amer. Math. Soc., 2007.
14. Pedro Duarte and Silvius Klein. Continuity, positivity and simplicity of the Lyapunov exponents for quasi-periodic cocycles. *Journal of the European Mathematical Society*, 21(7):2051–2106, 2019.
15. Pedro Duarte and Silvius Klein. Large deviations for products of random two dimensional matrices. *Communications in Mathematical Physics*, 375:2191–2257, 2020.
16. Pedro Duarte, Silvius Klein, et al. Lyapunov exponents of linear cocycles. *Atlantis Studies in Dynamical Systems*, 3, 2016.
17. Pedro Duarte, Silvius Klein, and Mauricio Poletti. Hölder continuity of the Lyapunov exponents of linear cocycles over hyperbolic maps. *Mathematische Zeitschrift*, 302(4):2285–2325, 2022.
18. Pedro Duarte, Silvius Klein, and Manuel Santos. A random cocycle with non Hölder Lyapunov exponent. *Discrete & Continuous Dynamical Systems*, 39(8):4841, 2019.
19. Catalina Freijo and Pedro Duarte. Continuity of the Lyapunov exponents of non-invertible random cocycles with constant rank. *arXiv preprint arXiv:2210.14851*, 2022.
20. Catalina Freijo and Karina Marin. Continuity of Lyapunov exponents for non-uniformly fiber-bunched cocycles. *Ergodic Theory and Dynamical Systems*, 41(12):3740–3767, 2021.
21. H. Furstenberg. Non-commuting random products. *Trans. Amer. Math. Soc.*, 108:377–428, 1963.
22. H. Furstenberg and H. Kesten. Products of random matrices. *Ann. Math. Statist.*, 31:457–469, 1960.
23. H. Furstenberg and Yu. Kifer. Random matrix products and measures in projective spaces. *Israel J. Math*, 10:12–32, 1983.
24. Y. Kifer. *Ergodic Theory of Random Transformations*, volume 10 of *Progress in Probability and Statictic*. Birkhäuser, 1986.
25. Yuri Kifer. Perturbations of random matrix products. *Zeitschrift für Wahrscheinlichkeitstheorie und Verwandte Gebiete*, 61(1):83–95, 1982.
26. J. Kingman. The ergodic theory of subadditive stochastic processes. *J. Royal Statist. Soc.*, 30:499–510, 1968.

27. E. Malheiro and M. Viana. Lyapunov exponents of linear cocycles over Markov shifts. *Stoch. Dyn.*, 15:1550020, 27, 2015.
28. V. I. Oseledets". A multiplicative ergodic theorem: Lyapunov characteristic numbers for dynamical systems. *Trans. Moscow Math. Soc.*, 19:197–231, 1968.
29. É. Le Page. Régularité du plus grand exposant caractéristique des produits de matrices aléatoires indépendantes et applications. *Ann. Inst. H. Poincaré Probab. Statist.*, 25:109–142, 1989.
30. Y. Peres. Analytic dependence of Lyapunov exponents on transition probabilities. In *Lyapunov exponents (Oberwolfach, 1990)*, volume 1486 of *Lecture Notes in Math.*, pages 64–80. Springer-Verlag, 1991.
31. David Ruelle. Analycity properties of the characteristic exponents of random matrix products. *Advances in mathematics*, 32(1):68–80, 1979.
32. Barry Simon and Michael Taylor. Harmonic analysis on SL(2,R) and smoothness of the density of states in the one-dimensional Anderson model. *Communications in mathematical physics*, 101:1–19, 1985.
33. A. Sánchez and M. Viana. Lyapunov exponents of probability distributions with non-compact support. Preprint https://arxiv.org/pdf/1810.03061.pdf.
34. E. Y. Tall and M. Viana. Moduli of continuity of Lyapunov exponents for random GL(2)-cocycles. *Trans. Amer. Math. Soc*, 373:1343–1383, 2020.
35. Marcelo Viana. *Lectures on Lyapunov exponents*, volume 145. Cambridge University Press, 2014.
36. Marcelo Viana. (dis) continuity of Lyapunov exponents. *Ergodic Theory and Dynamical Systems*, 40(3):577–611, 2020.
37. Marcelo Viana and Jiagang Yang. Continuity of Lyapunov exponents in the c^0 topology. *Israel Journal of Mathematics*, 229(1):461–485, 2019.
38. Yiqian Wang and Jiangong You. Quasi-periodic Schrödinger cocycles with positive Lyapunov exponent are not open in the smooth topology. *arXiv preprint arXiv:1501.05380*, 2015.

Some Questions and Remarks on Lyapunov Irregular Behavior for Linear Cocycles

Paulo Varandas

Abstract In this note we consider continuous cocycles over the shift taking values on a Lie subgroup $\mathcal{G}$ of $SL(2,\mathbb{R})$. While the celebrated Oseledets theorem guarantees that the Lyapunov exponents are defined in a total probability set the set of Lyapunov irregular points is well known to be topologically large for real valued cocycles. Focusing on the context of locally constant cocycles, we discuss a relation between the absence of Lyapunov irregular points and a rigidity phenomenon, namely that the cocycle is cohomologous to a constant matrix cocycle.

Keywords Cohomological equation · Rigidity · Lyapunov exponents

2020 Mathematics Subject Classification Primary 37D20, 37C20, 37C15; Secondary 37H15

1 Introduction

In the seminal papers [18, 19], Livšic considered the problem of obtaining solutions of the cohomological equation

$$A(x) = C(f(x))B(x)C(x)^{-1} \tag{1.1}$$

over hyperbolic dynamical systems $f : M \to M$ and for cocycles $A, B : M \to \mathcal{G}$ taking values on a Lie group $\mathcal{G}$ that is either $\mathbb{R}$ or an abelian group. Livšic focused on the following problems: (1) existence of continuous solutions C from the existence of a a priori measurable solution for (1.1) (with respect to some fully supported probability measure), and (2) obtaining necessary and sufficient conditions for the existence of solutions for (1.1), based on the evaluation of the cocycle at periodic

P. Varandas (✉)
Departamento de Matemática, Universidade Federal da Bahia, Salvador, Brazil
e-mail: paulo.varandas@ufba.br

© The Author(s), under exclusive license to Springer Nature Switzerland AG 2023

J. L. Dias et al. (eds.), *New Trends in Lyapunov Exponents*, CIM Series in Mathematical Sciences, https://doi.org/10.1007/978-3-031-41316-2_8

points. In the case of real valued Hölder cocycles both problems have a positive answer, and it is well known that the solution of the cohomological equation exists and is unique up to an additive constant. The context of Hölder continuous cocycles over hyperbolic systems taking values on some abelian group has been extensively studied, specially with the focus of establishing sufficient conditions for cohomology in terms of the periodic data and on studying the regularity of the conjugacy [19]. Actually we have the following consequence from [19] (see also [6, Proposition 4.5]):

Theorem 1.1 (Livšic) *If $\Sigma_\kappa = \{1, 2, \dots, \kappa\}^{\mathbb{Z}}$, $\sigma : \Sigma_\kappa \to \Sigma_\kappa$ is the shift map and $A : \Sigma_\kappa \to \mathbb{R}$ is a Hölder continuous function then the following are equivalent:*

(a) A is cohomologous to a constant (i.e. there exists $c \in \mathbb{R}$ and a Hölder continuous function $u : \Sigma_\kappa \to \mathbb{R}$ so that $A = u \circ \sigma - u + c$);
(b) if $x, y \in \Sigma_\kappa$ are periodic points for σ, $\sigma^n(x) = x$ and $\sigma^m(y) = y$ then

$$\frac{1}{n}\sum_{j=0}^{n-1} A(\sigma^j(x)) = \frac{1}{m}\sum_{j=0}^{m-1} A(\sigma^j(y));$$

(c) the Birkhoff averages $\left(\frac{1}{n}\sum_{j=0}^{n-1} A(\sigma^j(\cdot))\right)_{n\geqslant 1}$ converge uniformly to a constant.

The previous theorem conveys a rigidity phenomenon, saying that Birkhoff averages of Hölder continuous observables are everywhere convergent if and only if these are cohomologous to a constant. The Livšic theorem also finds applications that go beyond the symbolic context, in the context of Anosov diffeomorphisms. Recall that $f \in \mathrm{Diff}^1(M)$ is an Anosov diffeomorphism if there exists a Df-invariant splitting $TM = E^s \oplus E^u$, and constants $C > 0$ and $\lambda \in (0, 1)$ such that

$$\|Df^n(x)\mid_{E^s_x}\| \leqslant C\lambda^n \qquad \text{and} \qquad \|Df^{-n}(x)\mid_{E^u_x}\| \leqslant C\lambda^n$$

for every $x \in M$ and $n \geqslant 0$. In the special case there exist Df-invariant splittings

$$E^s = E^s_1 \oplus E^s_2 \oplus \cdots \oplus E^s_{n_s} \qquad \text{and} \qquad E^u = E^u_1 \oplus E^u_2 \oplus \cdots \oplus E^u_{n_u}$$

in one-dimensional subspaces, Livšic theorem can be applied to deal separately with the Birkhoff averages of the real valued Hölder continuous observables

$$A_{s,i}(x) = \log \|Df(x)\mid_{E^s_{i,x}}\| \text{ and } A_{u,j}(x) = \log \|Df(x)\mid_{E^u_{j,x}}\|,$$

where $1 \leqslant i \leqslant n_s$ and $1 \leqslant j \leqslant n_u$. In this special context the previous Birkhoff averages coincide with the Lyapunov exponents of the Anosov diffeomorphism f (as defined by Oseledets [20]) and, in particular, the Lyapunov exponents are everywhere well defined if and only if all periodic points have the same Lyapunov exponents. We should mention that a much finer rigidity result has been obtained recently by de la Llave and Micena [10]: every $C^{1+\alpha}$-Anosov diffeomorphism f on the torus $\mathbb{T}^d$ $(d \geqslant 2)$ whose Lyapunov exponents are everywhere well defined and that is C^1-close to an Anosov automorphism f_0 diagonalizable over $\mathbb{R}$, irreducible over $\mathbb{Q}$, whose eigenvalues are all distinct in modulus is C^{1+}-conjugate to f_0 (cf. [10, Theorem A] for more details).

An extension of Livšic theorem to the context of non-abelian groups $\mathcal{G}$ is more delicate and presents conceptual obstructions even in the context of matrix valued cocycles. In order to be more precise let us recall some concepts. A *linear cocycle* (over the shift σ) taking values on $\mathcal{G}$ is a pair (A, σ) where $A\colon \Sigma_\kappa \to \mathcal{G}$ is a matrix-valued map. The space $C^\nu(\Sigma_\kappa, \mathcal{G})$ of ν-Hölder continuous maps $A\colon \Sigma_\kappa \to \mathcal{G}$ endowed with the norm $\|A\|_\nu := \|A\|_\infty + \sup_{x\neq y} \frac{\|A(x)-A(y)\|}{d(x,y)^\nu}$ is a Banach space. One can associate to $A \in C^\nu(\Sigma_\kappa, \mathcal{G})$ the skew-product

$$\begin{aligned} F_A\colon \Sigma_\kappa \times \mathbb{K}^d &\longrightarrow \quad \Sigma_\kappa \times \mathbb{K}^d \\ (x, v) \quad &\mapsto \quad (\sigma(x), A(x)\cdot v). \end{aligned} \tag{1.2}$$

For each $n \geqslant 1$ and $(x, v) \in \Sigma_\kappa \times \mathbb{K}^d$ one can write $F_A^n(x, v) = (\sigma^n(x), A^n(x)v)$, where

$$A^n(x) = \begin{cases} A(\sigma^{n-1}(x)) \cdots A(\sigma(x))A(x), & \text{if } n \geqslant 0 \\ A(\sigma^n(x))^{-1} \dots A(\sigma^{-1}(x))^{-1}, & \text{otherwise.} \end{cases}$$

In this context, Kingman's subadditive ergodic theorem ensures that for every σ-invariant probability measure μ so that $\int \log \|A^{\pm 1}\|\, d\mu < \infty$ there exists a full μ-measure subset $\Sigma_\mu \subset \Sigma_\kappa$ so that the limits

$$\chi^+(x) = \lim_{n\to\pm\infty} \frac{1}{n} \log \|A^n(x)\| \quad \text{and} \quad \chi^-(x) = \lim_{n\to\pm\infty} \frac{1}{n} \log \|(A^n(x))^{-1}\|^{-1} \tag{1.3}$$

are well defined for every $x \in \Sigma_\mu$. These are the called the largest and the smallest Lyapunov exponents of the cocycle A at the point x, respectively. More generally, *Lyapunov exponents* are defined using Oseledets' theorem (cf. Sect. 3.2), and points for which the Lyapunov exponents are well defined are called Lyapunov regular.

The *Lyapunov irregular* set is formed by all points for which the Lyapunov exponents are not well defined. The Lyapunov irregular set is a zero measure set for all invariant measures but nonetheless it is often topologically large (see e.g. [12] and references therein for precise statements). We say that two cocycles $A, B \in C^\nu(\Sigma_\kappa, \mathcal{G})$ are *measurable cohomologous* (resp. *Hölder cohomologous*) if there exists a measurable map (resp. Hölder continuous map) $C : M \to \mathcal{G}$ such that $A(x) = C(\sigma(x))B(x)C(x)^{-1}$ for all $x \in M$. In general there are several obstructions to obtain a counterpart of Livšic theorem (Theorem 1.1) for cocycles taking values on non-abelian Lie groups. Even in the context of matrix valued cocycles over the shift: (1) there exists a pair of $GL(d, \mathbb{R})$ valued linear cocycles that are measurable cohomologous but not Hölder cohomologous, and (2) there exists pair of non-cohomologous $GL(d, \mathbb{R})$ valued linear cocycles whose matrices corresponding to periodic points have the same eigenvalues (even among important classes of fiber-bunched cocycles). This shows that Livšic theorem does not admit a simple extension to the context of non-abelian matrix cocycles. We refer the reader to [2, 16, 21, 23, 24, 27] and references therein for recent contributions on this topic, which remains not completely understood.

The main goal of this note is to contribute further to the study of Lyapunov irregular behavior for linear cocycles, namely the relation between irregular sets and rigidity. Let us first introduce some notation. Given an arbitrary Lie group $\mathcal{G}$ consider the sets

$$\mathcal{A}_1(\mathcal{G}) = \Big\{A \in C^\nu(\Sigma_\kappa, \mathcal{G}) \colon \text{A is cohomologous to a constant matrix cocycle}\Big\},$$

$$\mathcal{A}_2(\mathcal{G}) = \Big\{A \in C^\nu(\Sigma_\kappa, \mathcal{G}) \colon \text{the set of Lyapunov irregular points is empty}\Big\},$$

and

$$\mathcal{A}_3(\mathcal{G}) = \Big\{A \in C^\nu(\Sigma_\kappa, \mathcal{G}) \colon \text{all Lyapunov exponents of A at periodic points coincide}\Big\}.$$

In general

$$\mathcal{A}_1(\mathcal{G}) \subseteq \mathcal{A}_2(\mathcal{G}) \subseteq \mathcal{A}_3(\mathcal{G}) \subseteq C^\nu(\Sigma_\kappa, \mathcal{G}) \tag{1.4}$$

(the first inclusion is immediate and second inclusion being a consequence of Proposition 3.1 below). In the special case that either $\mathcal{G} = \mathbb{R}$ or $\mathcal{G}$ is an abelian Lie group, Livšic theorem implies that the previous three sets coincide and that such sets of rigidity form a meager subset of $C^\nu(\Sigma_\kappa, \mathbb{R})$. The equality between the three sets in (1.4) is no longer true in the context of arbitrary linear cocycles, even if $\mathcal{G} = \mathrm{SL}(2, \mathbb{R})$ [22]. This discussion suggests to look for a better understanding of the sets $\mathcal{A}_2(\mathcal{G}) \setminus \mathcal{A}_1(\mathcal{G})$ and $\mathcal{A}_3(\mathcal{G}) \setminus \mathcal{A}_2(\mathcal{G})$. More precisely, it is natural to look for conditions which ensure that the inclusions in (1.4) are proper and each of the sets $\mathcal{A}_i(\mathcal{G})$ is a meager subset of $\mathcal{A}_{i+1}(\mathcal{G})$.

In this context of linear cocycles over the shift, it is known under great generality that the set of Lyapunov irregular points is either empty or it is a Baire generic subset of the shift space (see e.g. [7, 12, 26] and references therein). In this note we discuss a relation between the absence of Lyapunov irregular points and a rigidity phenomenon, namely that the cocycle is cohomologous to a constant matrix cocycle. In the context of locally constant cocycles, such relation turns out to depend both on the Lie subgroup $\mathcal{G}$ where the cocycle takes values and on its locus of hyperbolicity/elipticity (cf. proof of Theorem 2.1). It is specifically due to the last dependence, namely the description of hyperbolic and elliptic cocycles in the space of cocycles (using [1, 29]), that one restricts our analysis to the set $C^0_{\text{loc}}(\Sigma_\kappa, \mathbb{R})$ of locally constant cocycles. Moreover, if the group $\mathcal{G}$ does not contain hyperbolic matrices then all points are Lyapunov regular for every cocycle $A : \Sigma_\kappa \to \mathcal{G}$ and the space of cocycles cohomologous to a constant matrix cocycle is a closed subset with empty interior in $C^0_{\text{loc}}(\Sigma_\kappa, \mathcal{G})$. Finally, if the group $\mathcal{G}$ contains hyperbolic matrices then the Lyapunov irregular set is non-empty if and only if there exist periodic points whose Lyapunov spectrum differs. We refer the reader to Theorem 2.1 for the precise statements.

The paper is organized as follows. In Sect. 2 we define the necessary concepts in order to state the main results. In Sect. 3 we recall and prove some preliminary results on dominated splittings, description of the space $C^0_{\text{loc}}(\Sigma_\kappa, \mathcal{G})$, and sufficient conditions for the existence of Lyapunov irregular behavior. The proof of the main result (Theorem 4.1) is given in Sect. 4.1.

2 Statement of the Main Results

In order to state our main results one needs to recall some necessary concepts. The *cohomology class* $\mathcal{C}(A)$ of a cocycle $A \in C^\nu(\Sigma_\kappa, \mathcal{G})$ is defined as the set of ν-Hölder continuous cocycles B that are ν-Hölder cohomologous to A. Given a finite collection of matrices $\mathcal{M} = \{A_1, \ldots, A_\kappa\} \subset \text{SL}(2, \mathbb{R})$, let $A \in C^0_{loc}(\Sigma_\kappa, \text{SL}(2, \mathbb{R}))$ denote the locally constant linear cocycle over the shift σ determined by $\mathcal{M}$ (these are ν-Hölder continuous). Defining $\mathcal{A}_{i,\text{loc}}(\mathcal{G}) = \mathcal{A}_i(\mathcal{G}) \cap C^0_{\text{loc}}(\Sigma_\kappa, \mathcal{G})$ for each $i = 1, 2, 3$ we have the inclusion

$$\mathcal{A}_{1,\text{loc}}(\mathcal{G}) \subseteq \mathcal{A}_{2,\text{loc}}(\mathcal{G}) \subseteq \mathcal{A}_{3,\text{loc}}(\mathcal{G}) \subseteq C^0_{\text{loc}}(\Sigma_\kappa, \mathcal{G}) \tag{2.1}$$

A straightforward computation shows that if $A \in C^0_{loc}(\Sigma_\kappa, \text{SL}(2, \mathbb{R}))$ is cohomologous to a constant matrix cocycle then there exists $A_0 \in \text{SL}(2, \mathbb{R})$ so that each matrix in $\mathcal{M}$ is linearly conjugate to A_0. The following result sheds some light on the nature of the inclusions of the sets appearing in Eq. (2.1) and, ultimately, over the inclusions in (1.4). The fact of having a strict inclusion or an equality involves both the Lie subgroup $\mathcal{G}$ and the space $\mathcal{H} \subset C^0_{loc}(\Sigma_\kappa, \text{SL}(2, \mathbb{R}))$ of hyperbolic cocycles (see Sect. 3.1 for the definition).

Theorem 2.1 *Let $\mathcal{G}$ be a Lie subgroup of $SL(2, \mathbb{R})$.*

(1) If $\mathcal{G}$ contains no hyperbolic matrix then:

(a) $\mathcal{A}_{2,loc}(\mathcal{G}) = \mathcal{A}_{3,loc}(\mathcal{G}) = C^0_{loc}(\Sigma_\kappa, \mathcal{G})$;
(b) $\mathcal{A}_{1,loc}(\mathcal{G})$ is a meager subset of $C^0_{loc}(\Sigma_\kappa, \mathcal{G})$.

(2) If $\mathcal{G}$ contains a hyperbolic matrix and $\kappa = 2$ then:

(a) $\mathcal{A}_{1,loc}(\mathcal{G}) \subsetneq \mathcal{A}_{2,loc}(\mathcal{G}) = \mathcal{A}_{3,loc}(\mathcal{G}) \subsetneq C^0_{loc}(\Sigma_\kappa, \mathcal{G})$;

(b) $\mathcal{A}_{1,loc}(\mathcal{G}) \cap \mathcal{H} = \mathcal{A}_{2,loc}(\mathcal{G}) \cap \mathcal{H}$;
(c) $\mathcal{A}_{1,loc}(\mathcal{G}) \cap \mathcal{H}^c$ *is a meager subset of* $\mathcal{A}_{2,loc}(\mathcal{G}) \cap \mathcal{H}^c$;
(d) $\mathcal{A}_{3,loc}(\mathcal{G})$ *is a meager subset of* $C^0_{loc}(\Sigma_\kappa, \mathcal{G})$.

Some remarks are in order. Items (1) and (2) present two different pictures according to the subgroup $\mathcal{G}$. The assumption $\kappa = 2$ in item (2) is used solely to describe the cocycles in the boundary of the space of elliptic cocycles, a problem which is not completely understood whenever $\kappa > 2$ (see Sect. 3.1 for more details). Moreover, the previous dichotomy cannot be expressed in terms of the compactness of the subgroup. In fact, item (1) above encloses both compact and some non-compact subgroups of $SL(2, \mathbb{R})$. Moreover, there exist such linear cocycles whose Lyapunov exponents are everywhere defined but the cocycle is not cohomologous to a constant cocycle (see Examples 2.1 and 2.2 below).

Example 2.1 Let $\sigma : \{0, 1\}^{\mathbb{Z}} \to \{0, 1\}^{\mathbb{Z}}$ be the shift and let $SO(2, \mathbb{R}) \subset SL(2, \mathbb{R})$ denote the special orthogonal group, isomorphic to $\mathbb{R}/2\pi\mathbb{Z} \simeq \mathbb{S}^1$. Let $\alpha : \{0, 1\}^{\mathbb{Z}} \to \mathbb{R}$ be Hölder continuous which is not cohomologous to a constant and consider the cocycle $A \in C^\nu(\{0, 1\}^{\mathbb{Z}}, SO(2, \mathbb{R}))$ given by

$$A(x) = \begin{pmatrix} \cos\alpha(x) & -\sin\alpha(x) \\ \sin\alpha(x) & \cos\alpha(x) \end{pmatrix}. \tag{2.2}$$

As A takes values in the compact group formed by rotations, it is clear that all points have zero Lyapunov exponents and that every point in $\{0, 1\}^{\mathbb{Z}}$ is Lyapunov regular. We claim that the cocycle A is not cohomologous to a constant cocycle. Indeed, otherwise there would exist $A_0 \in SL(2, \mathbb{R})$ and $Q : \{0, 1\}^{\mathbb{Z}} \to SL(2, \mathbb{R})$ so that $A(x) = Q(\sigma(x))^{-1} A_0 Q(x)$ for every $x \in \{0, 1\}^{\mathbb{Z}}$. This implies that A_0 has only (generalized) eigenvalues of norm one, and that

$$A_0^n = A^n(x) = \begin{pmatrix} \cos\left(\sum_{j=0}^{n-1} \alpha(\sigma^j(x))\right) & -\sin\left(\sum_{j=0}^{n-1} \alpha(\sigma^j(x))\right) \\ \sin\left(\sum_{j=0}^{n-1} \alpha(\sigma^j(x))\right) & \cos\left(\sum_{j=0}^{n-1} \alpha(\sigma^j(x))\right) \end{pmatrix} \tag{2.3}$$

for every $x \in \{0, 1\}^{\mathbb{Z}}$ so that $\sigma^n(x) = x$. However, as α is not cohomologous to a constant Theorem 1.1 implies that there exist sequences $x, y \in \{0, 1\}^{\mathbb{Z}}$ so that $\sigma^n(x) = x$, $\sigma^m(y) = y$ but $A^{(mn)}(x) \neq A^{(mn)}(y)$, leading to a contradiction with (2.3).

Example 2.2 Let $\sigma : \{0, 1\}^{\mathbb{Z}} \to \{0, 1\}^{\mathbb{Z}}$ be the shift and let $\alpha : \{0, 1\}^{\mathbb{Z}} \to \mathbb{R}$ be Hölder continuous which is not cohomologous to a constant. Consider the cocycle $A \in C^\nu(\{0, 1\}^{\mathbb{Z}}, SL(2, \mathbb{R}))$ given by

$$A(x) = \begin{pmatrix} 1 & \alpha(x) \\ 0 & 1 \end{pmatrix}. \tag{2.4}$$

Even though A takes values in a non-compact group, it is clear that all points have only zero Lyapunov exponents and so every point in $\{0, 1\}^{\mathbb{Z}}$ is Lyapunov regular.

Observe also that, for each $n \geqslant 0$,

$$A^n(x) = \begin{pmatrix} 1 & \sum_{j=0}^{n-1} \alpha(\sigma^j(x)) \\ 0 & 1 \end{pmatrix}, \qquad x \in \{0,1\}^{\mathbb{Z}} \tag{2.5}$$

An argument identical to the one used in Example 2.1 shows that the cocycle A is not cohomologous to a constant cocycle.

Finally, we observe that Theorem 2.1 can be used to deal with locally constant cocycles admitting a finest invariant splitting which is continuous and formed by one-dimensional and two-dimensional subbundles. For simplicity, we illustrate this for cocycles having a dominated splitting in dimension three (we refer the reader to Sect. 3.1 for the definition of dominated splittings and uniformly hyperbolic cocycles).

Corollary 2.1 *Assume that $A \in C^0_{loc}(\Sigma_\kappa, SL(3,\mathbb{R}))$ admits an A-invariant dominated splitting $\Sigma_\kappa \times \mathbb{R}^3 = E \oplus F$, where $\dim E = 2$ and $\dim F = 1$. The cocycle A has no Lyapunov irregular points if and only if A is uniformly hyperbolic and one of the following properties hold:*

(1) A is cohomologous to a constant matrix cocycle with three real eigenvalues; or
(2) there exists $c \in \mathbb{R}$ so that

$$\lim_{n\to\infty} \frac{1}{n} \log \|A^n(x)v\| = c, \qquad \forall x \in \Sigma_\kappa, \quad \forall v \in E_x \setminus \{0\}.$$

3 Preliminaries

3.1 Hyperbolicity, Parabolicity and Elipticity

A matrix $A_0 \in SL(2,\mathbb{R})$ is called *hyperbolic* if $\mathrm{trace}(A_0) > 2$, *elliptic* if $\mathrm{trace}(A_0) < 2$ and it is called *parabolic* if $\mathrm{trace}(A_0) = 2$. It is clear that the set of hyperbolic matrices and the set of elliptic matrices are both open subsets of $SL(2,\mathbb{R})$, whose union is dense in $SL(2,\mathbb{R})$.

Given a continuous cocycle $A : \Sigma_\kappa \to \mathcal{G}$ we say that an A-invariant splitting $\Sigma_\kappa \times \mathbb{R}^d = E^1 \oplus E^2$ is *dominated* if there exists $C > 0$ and $\lambda \in (0,1)$ so that

$$\|A^n(x) \mid_{E^2_x} \|.\|(A^n(x) \mid_{E^1_x})^{-1}\| \leqslant C\lambda^n \text{ for every } n \geqslant 1 \text{ and } x \in \Sigma_\kappa.$$

We recall that dominated splittings are well known to depend continuously with the point.

We say that the cocycle $A : \Sigma_\kappa \to \mathcal{G}$ is *uniformly hyperbolic* if there exist an A-invariant splitting $\Sigma_\kappa \times \mathbb{R}^d = E^s \oplus E^u$ and constants $C > 0$ and $\lambda \in (0,1)$ so that

$$\|A^n(x)\mid_{E^s_x}\| \leqslant C\lambda^n \quad \text{and} \quad \|(A^n(x)\mid_{E^u_x})^{-1}\| \leqslant C\lambda^n$$

for every $x \in \Sigma_\kappa$ and $n \geqslant 0$. The space of hyperbolic cocycles will be denoted by $\mathcal{H}$. We refer the reader to [29] for other characterizations of uniform hyperbolicity in the context of $SL(2,\mathbb{R})$-cocycles which, in particular, is a C^0-open condition in $C^0(\Sigma_\kappa, SL(2,\mathbb{R}))$.

A linear cocycle $A : \Sigma_\kappa \to \mathcal{G}$ is *elliptic* if there exists a periodic point $x \in \Sigma_\kappa$ of period $k \geqslant 1$ so that $A^k(x)$ is an elliptic matrix. The space of elliptic cocycles will be denoted by $\mathcal{E}$. We have the following fine characterization of the boundary of the space of hyperbolic cocycles over the full shift on two symbols:

Theorem 3.1 ([1, Theorem 3.2 and 3.3]) *If* $\kappa = 2$ *then*

$$\partial\mathcal{H} = \partial\mathcal{E} = (\mathcal{H} \cup \mathcal{E})^c.$$

Moreover, if $(A_1, A_2) \in \mathcal{G} \times \mathcal{G}$ *generates a cocycle* $A \in C^0_{loc}(\Sigma_2, \mathcal{G})$ *which belongs to the boundary of* $\mathcal{H}$ *then at least one of the following possibilities hold:*

(1) there exists a periodic point $x \in \Sigma_2$ *of period* k *for* σ *such that* $A^k(x)$ *is parabolic,*
(2) the matrices A_1 *and* A_2 *are hyperbolic and either* $E^s_{A_1} = E^u_{A_2}$ *or* $E^u_{A_1} = E^s_{A_2}$.

One could ask what is the importance of having products of matrices with two generators in the previous theorem. Actually, there is a similar description of the boundary of the space of hyperbolic cocycles among locally constant cocycles obtained by random products of κ matrices ($\kappa \geqslant 2$). More precisely:

Theorem 3.2 ([1, Theorem 4.1]) *Let* $A_1, \ldots, A_\kappa$ *belong to the boundary of a connected component of* $\mathcal{H}$. *Then at least one of the following possibilities hold:*

(1) there exists a periodic point $x \in \Sigma_\kappa$ *of period* k *for* σ *such that* $A^k(x)$ *is parabolic, or*
(2) there exist periodic points x, y *of* σ *with periods* $m, n \geqslant 1$, *an integer* $\ell \geqslant 0$ *and an homoclinic point* $z \in W^u_{loc}(x) \cap \sigma^{-\ell}(W^s_{loc}(y))$ *such that* $A^m(x)$ *and* $A^n(x)$ *are hyperbolic matrices and* $A^\ell(z)E^u_{A^m(x)} = E^s_{A^n(y)}$.

Remark 3.1 It is known that $\overline{\mathcal{E}} = \mathcal{H}^c$ (here $\overline{\mathcal{E}}$ stands for the closure of $\mathcal{E}$ and $\mathcal{H}^c$ stands for the complement of the set $\mathcal{H}$) [29]. However, in case of cocycles determined by random products of more than two matrices the boundary of the sets $\mathcal{H}$ and $\mathcal{E}$ need not coincide (we refer the reader to [1, 9, 29] for more details).

A characterization of cocycles in the boundary $\partial\mathcal{E}$ is not so immediate. Let us elaborate more on this topic. The *joint spectral radius* of a finite collection of matrices $\mathcal{M} = \{A_1, \ldots, A_\kappa\} \in \mathrm{SL}(2, \mathbb{R})$ is defined as

$$\rho_+(\mathcal{M}) = \inf_{n\to\infty} \sup\{\|A_{i_n} \ldots A_{i_2} A_{i_1}\|^{\frac{1}{n}} : A_{i_j} \in \mathcal{M}\},$$

whereas its *lower spectral radius* is defined as

$$\rho_-(\mathcal{M}) = \inf_{n\to\infty} \inf\{\|A_{i_n} \ldots A_{i_2} A_{i_1}\|^{\frac{1}{n}} : A_{i_j} \in \mathcal{M}\}.$$

Using Berger-Wang formula [4] it is often convenient to write the joint spectral radius using the spectral radius of the corresponding products of matrices by

$$\rho_+(\mathcal{M}) = \inf_{n\to\infty} \sup\{\rho(A_{i_n} \ldots A_{i_2} A_{i_1})^{\frac{1}{n}} : A_{i_j} \in \mathcal{M}\}, \tag{3.1}$$

By some abuse of notation, for any given $A \in C^0_{loc}(\Sigma_\kappa, \mathrm{SL}(2, \mathbb{R}))$ we write $\rho_+(A)$ (resp. $\rho_-(A)$) to denote the joint spectral radius (resp. lower spectral radius) of the collection of matrices which define the locally constant cocycle. The joint spectral radius as defined above is a continuous function [14], however the lower spectral is an upper-semicontinuous function (it is the infimum of continuous functions) which need not be continuous (see e.g. [5] and references therein). The next lemma gives a characterization of semigroups $\mathcal{S}$ which have no hyperbolic matrices in terms of the joint spectral radius.

Lemma 3.1 *Let $\mathcal{S} \subset SL(2, \mathbb{R})$ be the semigroup generated by the finite collection of matrices $A_1, \ldots, A_\kappa \in SL(2, \mathbb{R})$ and assume that the linear cocycle $A : \Sigma_\kappa \to SL(2, \mathbb{R})$ is elliptic. Then $\mathcal{S}$ contains no hyperbolic matrix if and only if $\rho(\mathcal{S}) = 1$. Moreover, if $\rho(\mathcal{S}) = 1$ then all points in Σ_κ are Lyapunov regular.*

Proof First notice that $\rho(\mathcal{S}) \geqslant 1$, as a consequence of the fact that $\|A_i\| \geqslant 1$ for every $1 \leqslant i \leqslant \kappa$. If $\rho(\mathcal{S}) = 1$ then it follows from the definition that the spectral radius of A_i is one for every $1 \leqslant i \leqslant \kappa$, and so $\sigma(A_i) \subset \mathbb{S}^1$ (here $\sigma(A_i) \subset \mathbb{C}$ denotes the spectrum of the matrix A_i). It is well known that $\sigma(A_i A_j) \subset \sigma(A_i)\sigma(A_j)$. Thus, $\sigma(A) \subset \mathbb{S}^1$ for all matrices $A \in \mathcal{S}$, which implies that the semigroup $\mathcal{S}$ does not contain hyperbolic matrices.

Conversely, if $\mathcal{S}$ does not contain hyperbolic matrices then the real canonical Jordan form theorem ensures that each generator A_i is either conjugated to a rotation (in case A_i has a pair of complex conjugate generalized eigenvalues), conjugated to $-Id$ (in case A_i has two eigenvectors associated to the eigenvalue -1) or conjugated to a nilpotent matrix (the later case corresponding to the case of a real eigenvalue with algebraic multiplicity two having a unique eigenvector). By assumption, the product of any of these matrices does not produce hyperbolic matrices. Thus, there exists a basis of $\mathbb{R}^2$ so that all matrices are nilpotent and can be written as upper triangular matrices, or either all matrices induce rotations (see e.g. proof of Theorem 2.1 at Sect. 4.1 for details). In both cases, the eigenvalues of matrices in $\mathcal{S}$ lie in the unit circle, hence the joint spectral radius is one.

Finally, if $\rho(\mathcal{S}) = 1$ then

$$0 \leqslant \liminf_{n\to\infty} \frac{1}{n} \log \|(A^n(x))^{-1}\|^{-1} \leqslant \liminf_{n\to\infty} \frac{1}{n} \log \|A^n(x)\| \leqslant \log \rho(\mathcal{S}) = 0$$

for every $x \in \Sigma_\kappa$, which proves that all points are Lyapunov regular and that zero is the only Lyapunov exponent. This proves the lemma. □

In the complementary direction, it would be interesting to use the lower spectral radius to detect elliptic behavior. Indeed, as $\rho_-(A) \geqslant 1$ for every $A \in \partial\mathcal{E}$, it would be interesting to have a positive answer for the following:

Question 1 Let $A \in C^0_{loc}(\Sigma_\kappa, \mathrm{SL}(2,\mathbb{R}))$ be the cocycle determined by a finite collection $\mathcal{M} = \{A_1, \ldots, A_\kappa\}$ of $\mathrm{SL}(2,\mathbb{R})$ matrices. If $A \in \partial\mathcal{E}$ does $\rho_-(\mathcal{M}) = 1$?

3.2 Oseledets' Theorem

Given a cocycle $A : \Sigma_\kappa \to \mathrm{SL}(d,\mathbb{R})$ satisfying the integrability assumption $\log \|A^{\pm 1}\| \in L^1(\mu)$, the Oseledets' theorem [20] ensures that for μ-a.e. $x \in \Sigma_\kappa$ there exist $1 \leqslant k(x) \leqslant d$, a splitting $\{x\} \times \mathbb{R}^d = E^1_x \oplus E^2_x \oplus \cdots \oplus E^{k(x)}_x$ (called the *Oseledets splitting*) varying measurably with x and so that $A(x)E^i_x = E^i_{\sigma(x)}$ for every $1 \leqslant i \leqslant k(x)$, and real numbers $\lambda_1(A,\sigma,x) > \lambda_2(A,\sigma,x) > \cdots > \lambda_{k(x)}(A,\sigma,x)$ (called the *Lyapunov exponents* associated to σ, A and x) such that

$$\lambda_i(A,\sigma,x) = \lim_{n\to\pm\infty} \frac{1}{n} \log \|A^n(x)v_i\|, \quad \forall v_i \in E^i_x \setminus \{\vec{0}\}. \tag{3.2}$$

If, in addition, μ is ergodic then both the number and the value of the Lyapunov exponents are almost everywhere constant, and the Lyapunov exponents are denoted simply by $\lambda_i(A,\mu)$. Moreover, the largest and smallest Lyapunov exponents of the cocycle with respect to an ergodic probability μ can be expressed, using Kingman's subadditive ergodic theorem, by

$$\lambda_+(A,\mu) := \lim_{n\to\infty} \frac{1}{n} \log \left\|A^n(x)\right\| \quad \text{and} \quad \lambda_-(A,\mu) := \lim_{n\to\infty} \frac{1}{n} \log \left\|A^n(x)^{-1}\right\|^{-1}, \tag{3.3}$$

respectively (see also the previous work by Furstenberg and Kesten [13] on the products of random matrices). Moreover, Oseledets' theorem ensures that for μ-almost every $x \in \Sigma_\kappa$

$$\sum_{i=1}^{d} \lambda_i(A,\mu) = \lim_{n\to\infty} \frac{1}{n} \log |\det A^n(x)| = 0. \tag{3.4}$$

We refer the reader to [11, 28] for two excellent monographs on Lyapunov exponents for linear cocycles and related questions.

3.3 GL(2, ℝ)-Valued Cocycles

The Lyapunov exponents of an SL(2, ℝ)-cocycle A have a natural symmetry. Indeed, it follows from Osedelets' theorem (recall (3.4)) that if μ is a σ-invariant and ergodic probability measure then it has two Lyapunov exponents $\chi^-(A,\mu) \leqslant 0 \leqslant \chi^-(A,\mu)$, where $\chi^-(\mu) = -\chi^+(\mu)$. Nevertheless, any GL(2, ℝ)-valued cocycle $B : \Sigma_\kappa \to \mathrm{GL}(2,\mathbb{R})$ can be written as

$$B(x) = \det B(x) \cdot A(x) \quad \text{where} \quad A(x) = \frac{B(x)}{\det B(x)} \tag{3.5}$$

is an SL(2, ℝ)-valued cocycle. Thus, the Lyapunov exponents of B and A are related by the simple expressions

$$\chi^\pm(B,\mu) = \chi^\pm(A,\mu) + \int \log|\det B(x)|\, d\mu.$$

Furthermore, as the integral in the right hand-side above corresponds to the sum of the Lyapunov exponents, a point $x \in \Sigma_\kappa$ is Lyapunov irregular with respect to the linear cocycle B if it is either a Birkhoff irregular point for the real valued observable $\varphi : \Sigma_\kappa \to \mathbb{R}$ given by $\varphi(x) = \log|\det B(x)|$, or a Lyapunov irregular point for the SL(2, ℝ)-valued normalized cocycle A defined in (3.5).

3.4 Lyapunov Irregular Points

In view of Kingman's subadditive ergodic theorem, the top Lyapunov exponent in (3.3) can be computed as an almost everywhere limit of a sub-additive family of continuous functions. Recall that a sequence $\Phi = (\varphi_n)_n$ of real valued functions is *sub-additive* if $\varphi_{m+n} \leqslant \varphi_m + \varphi_n \circ \sigma^n$ for every $m, n \geqslant 0$. Carvalho and Varandas [7] proved that if $\Phi = (\varphi_n)_n$ is a sequence of sub-additive continuous functions and there exist two dense subsets $D_1, D_2 \subset \Sigma_\kappa$ so that

$$\inf_{n\geqslant 1} \frac{1}{n}\varphi_n(x) \leqslant \alpha < \beta \leqslant \inf_{n\geqslant 1} \frac{1}{n}\varphi_n(y) \tag{3.6}$$

for every $x \in D_1$ and $y \in D_2$ then the set

$$\left\{x \in \Sigma_\kappa : \lim_{n\to\infty} \frac{1}{n}\varphi_n(x) \text{ does not exist}\right\}$$

of Φ-*irregular points* is a Baire generic subset of Σ_κ (see also [8] for a more general result). Notice that in the case of the shift every point has a dense set of preimages. In consequence, we conclude that if there exist periodic points $x, y \in \Sigma_\kappa$ so that (3.6) holds then the set of Φ-irregular points is non-empty. In the context of a linear cocycle $A : \Sigma_\kappa \to \mathrm{SL}(2, \mathbb{R})$ it is natural to consider the family $\Phi = (\varphi_n)_{n\geqslant 1}$ of continuous and sub-additive functions as $\varphi_n(x) = \log \|A^n(x)\|$, $x \in \Sigma_\kappa$. As Lyapunov exponents sum zero for $\mathrm{SL}(2, \mathbb{R})$-cocycles, using [7, Theorem B] one derives the following immediate application:

Proposition 3.1 *Let $A : \Sigma_\kappa \to SL(2, \mathbb{R})$ be a (Hölder) continuous hyperbolic cocycle over the shift $\sigma : \Sigma_\kappa \to \Sigma_\kappa$. If there exist two periodic points whose Lyapunov exponents differ then the Lyapunov irregular set is a Baire generic subset of Σ_κ.*

We recall that Lemma 3.1 ensures that all points in the shift space are Lyapunov regular for a locally constant elliptic cocycle, whose associated semigroup contains no hyperbolic matrices. However, such elliptic cocycles need not be cohomologous to a constant matrix cocycle (cf. Example 2.1).

In case the semigroup $\mathcal{S}$ is elliptic, non-compact and contains hyperbolic matrices, the cocycle A is such that there exist σ-invariant probability measures μ_1, μ_2 so that $\chi^-(A, \mu_1) = 0 = \chi^+(A, \mu_1)$ and $\chi^-(A, \mu_2) < 0 < \chi^+(A, \mu_2)$, a condition which ensures that there exists a Baire generic subset of Σ_κ formed by Lyapunov irregular points (recall Proposition 3.1), and so one has the following:

Corollary 3.1 *Assume that the semigroup $\mathcal{S} \subset SL(2, \mathbb{R})$ generated by the finite collection of matrices $A_1, \ldots, A_\kappa \in SL(2, \mathbb{R})$ contains both hyperbolic and elliptic matrices. Then the corresponding locally constant cocycle $A : \{1, 2, \ldots, \kappa\}^{\mathbb{Z}} \to SL(2, \mathbb{R})$ is not cohomologous to a constant matrix cocycle.*

To finalize this subsection we illustrate how the Lyapunov irregular set can be used to describe shapes in refinement schemes for simplicial meshes in computational dynamics (see e.g. [25] and references therein). We focus in the geometric context of barycentric subdivision of triangles. In [3], Bárány, Beardon and Carne considered the successive barycentric subdivisions of a triangle $\mathcal{T}$. The nth-generation of this procedure consists of 6^n small triangles $\mathcal{T}_n$ obtained by barycentric subdivision from the ones of the $(n-1)^{\mathrm{th}}$-generation. Since all possible shapes of triangles can be obtained by permutation of the vertices of each of the six smaller triangles, there exists an itinerary map assigning to each triangle $T \in \mathcal{T}_n$ a word $\underline{\omega} = (\omega_1, \omega_2, \ldots, \omega_n) \in \{1, 2, 3, \ldots, 18\}^n$ (cf. [3]). Similarly, for each word $\underline{\omega} \in \{1, 2, 3, \ldots, 18\}^n$ one denotes by $\mathcal{T}_{\underline{\omega}} \in \mathcal{T}_n$ the triangle determined by $\underline{\omega}$. As it will be convenient to use the full shift Σ_{18} on eighteen symbols, for each $\underline{\omega} \in \Sigma_{18}$ we denote by $\underline{\omega}\mid_n \in \{1, 2, 3, \ldots, 18\}^n$ its length-n truncation.

It is known that successive barycentric subdivisions of a triangle contain triangles which (to within a similarity) approximate arbitrarily closely any given triangle, and that the shape of almost every randomly chosen triangle will converge to a flat triangle (cf. [3, Theorems 1 and 2]). Here we prove the following:

Corollary 3.2 *Given the successive barycentric subdivisions of a triangle, there exists a Baire generic subset of the space Σ_{18} so that the shape of the triangles $\mathcal{T}_{\underline{\omega}|_n}$ does not converge.*

Proof We use the characterization of the possible shapes of successive barycentric subdivisions of the triangle $\mathcal{T}$ obtained in [3]. Indeed, the proof of Theorems 1 and 2 in [3] assigns to each triangle $\mathcal{T}$ another triangle with the shape (a, b, c) in the hyperbolic plane $\mathbb{H}^+ = \{z \in \mathbb{C}\colon Im(z) > 0\}$ and shows that all the triangles in $\mathcal{T}_n$ are of the form $g_{i_n} \circ \cdots \circ g_{i_2} \circ g_{i_1}(\mathcal{T})$, for some Möebius map

$$\underline{g} = g_{i_n} \circ \cdots \circ g_{i_2} \circ g_{i_1} \tag{3.7}$$

where each g_{i_j} belongs to a fixed collection of eighteen Möebius maps in $\mathrm{Aut}(\mathbb{H}^+)$, including the Möebius maps

$$g(z) = \frac{2z+2}{3} \quad \text{and} \quad h(z) = \frac{z-2}{3z}.$$

The latter induce the matrices

$$A = \begin{pmatrix} 2/\sqrt{6} & 2/\sqrt{6} \\ 0 & 3/\sqrt{6} \end{pmatrix}, \quad B = \begin{pmatrix} 1/\sqrt{6} & -2/\sqrt{6} \\ 3/\sqrt{6} & 0 \end{pmatrix} \in \mathrm{SL}(2, \mathbb{R})$$

that do not commute, A is hyperbolic and B is an irrational rotation (cf. [3, pp. 167–168]). Moreover, if $G \in \mathrm{SL}(2, \mathbb{R})$ represents the Möebius map in (3.7) and $d_H(\cdot, \cdot)$ denotes the hyperbolic metric in $\mathbb{H}^+$ then

$$\left\| G \cdot \begin{pmatrix} \sigma \\ 1 \end{pmatrix} \right\|^2 = 2 \cdot \cos h(d_H(\underline{g}(\sigma), \mathbf{i})) \cdot Im(\sigma) \tag{3.8}$$

(cf. [3, p. 170]). Moreover, every triangle (a, b, c) can be written, up to a Möebius map, in the form $(0, 1, \sigma)$, its shape is determined by the images of the previous composition of Möebius maps. In this way, (3.8) relates the $g_{i_n} \circ \cdots \circ g_{i_2} \circ g_{i_1}(\mathcal{T})$ with the norm of the image of the vector $(\sigma, 1)$ by the matrix G and guarantees that the non-existence of the Lyapunov exponent is equivalent to the non-convergence of the shapes for this sequence of triangles. By Proposition 3.1, there exists a Baire generic subset $\mathcal{R} \subset \Sigma_{18}$ such that the shapes of the nested sequence of triangles $(\mathcal{T}_{\underline{\omega}|_n})_n$ determined by an element $\underline{\omega} \in \mathcal{R}$ do not converge. This finishes the proof of the corollary. □

Altogether, Theorem 1 in [3] and Corollary 3.2 ensure that, from the topological viewpoint, the shape of typical triangles obtained by successive barycentric subdivisions does not converge. However, the expression (3.8) just relates the shape of triangles with the norm of certain vectors, not with the vectors themselves. In view of [12] one can expect the following question to have a positive answer:

Question 2 Given the successive barycentric subdivisions of a triangle, does there exists a Baire generic subset of nested sequences of triangles whose refinements produce arbitrary shapes? More precisely, if $\mathcal{S}$ denotes the space of shapes of triangles in $\mathbb{H}^+$, does there exist a Baire generic subset $\mathcal{R} \subset \Sigma_{18}$ so that for each $\underline{\omega} \in \mathcal{R}$ the shape of the triangles $\mathcal{T}_{\underline{\omega}|_n}$ is dense in $\mathcal{S}$?

4 Proofs

4.1 *Proof of Theorem 2.1*

Let $\mathcal{G}$ be a Lie subgroup of SL(2, $\mathbb{R}$). Given a finite collection of matrices $\{A_1, \ldots, A_\kappa\}$ in $\mathcal{G}$ let $A \in C^0_{loc}(\Sigma_\kappa, \mathrm{SL}(2, \mathbb{R}))$ denote the associated locally constant linear cocycle, and let $\mathcal{S}$ denote the semigroup generated by them. As $|\det(A)| = 1$ for every $A_0 \in \mathcal{G}$, the Oseledets' theorem implies that the Lyapunov exponents are symmetric: if μ is a σ-invariant and ergodic probability measure then $\lambda^-(A, \mu) = -\lambda^+(A, \mu)$.

By the Iwasawa decomposition [15], every matrix $A_0 \in \mathrm{SL}(2, \mathbb{R})$ can be written as a product of an elliptic, a hyperbolic and a nilpotent matrix. If $\mathcal{G}$ contains no hyperbolic matrix then every matrix of $\mathcal{G}$ is either elliptic, nilpotent or a product of both. In particular the eigenvalues of every matrix in $\mathcal{G}$ are contained in the unit circle and, by the Berger-Wang formula for the joint spectral radius (recall (3.1)), one concludes that $\rho_+(A_1, \ldots, A_\kappa) = 1$. In consequence

$$\lim_{n\to\infty} \frac{1}{n} \log \|A^n(x)^{-1}\|^{-1} = \lim_{n\to\infty} \frac{1}{n} \log \|A^n(x)\| = 0 \quad \text{for every } x \in \Sigma_\kappa,$$

and so $\mathcal{A}_{2,\mathrm{loc}}(\mathcal{G}) = \mathcal{A}_{3,\mathrm{loc}}(\mathcal{G}) = C^0_{\mathrm{loc}}(\Sigma_\kappa, \mathcal{G})$. This proves item 1(a).

Now, if $A \in \mathcal{A}_{1,\mathrm{loc}}(\mathcal{G})$ then there exists $Q : \Sigma_\kappa \to \mathcal{G}$ and $A_0 \in \mathcal{G}$ so that $A(x) = Q(\sigma x)^{-1} A_0 Q(x)$ for every $x \in \Sigma_\kappa$. This implies that every matrix in the semigroup $\mathcal{S}$ is conjugate to an integer power of A_0. We claim that all matrices A_i ($1 \leqslant i \leqslant \kappa$) are either rotation matrices, or there exists a basis where these matrices can be represented by upper triangular nilpotent matrices. Indeed, as these matrices are pairwise conjugate, if A_1 is a rotation matrix (resp. nilpotent matrix) then the same holds for every A_i with $2 \leqslant i \leqslant \kappa$. Furthermore, in case A_1 is nilpotent one can choose a basis where A_1 can be represented by the upper triangular nilpotent matrix

$$\begin{pmatrix} 1 & a \\ 0 & 1 \end{pmatrix} \qquad \text{for some } a < 0$$

let A_i be represented in that basis by the matrix

$$\begin{pmatrix} u_i & v_i \\ w_i & z_i \end{pmatrix} \qquad \text{where } u_i + z_i = 2.$$

As $A_1 A_i$ is also a nilpotent matrix, or equivalently trace$(A_1 A_i) = 2$, an easy computation shows that $w_i = 0$ and $u_i = z_i = 1$. Altogether, this shows that A is either a cocycle formed by rotations in the form (2.2) or a cocycle formed by nilpotent matrices of the form (2.4). In either cases, the space of cocycles which are cohomologous to a constant matrix cocycle is a meager subset of $C^0_{\text{loc}}(\Sigma_\kappa, \mathcal{G})$ (recall Examples 2.1 and 2.2, and notice that Theorem 1.1 guarantees that the space of real valued observables cohomologous to a constant form a meager subset of $C^0(\Sigma_\kappa, \mathbb{R})$). This completes the proof of item 1 (b).

In the remainder of the proof we assume that $\mathcal{G}$ contains some hyperbolic matrix and $\kappa = 2$. Thus, the cocycles are generated by two matrices $A_1, A_2 \in \mathcal{G}$. We start by noticing that if $A \in \mathcal{A}_{3,\text{loc}}(\mathcal{G})$ then

$$A_1 \text{ and } A_2 \text{ have the same eigenvalues in absolute value.} \qquad (\star)$$

It is clear that the set of pairs (A_1, A_2) satisfying $(\star)$ form a closed subset of $\mathcal{G} \times \mathcal{G}$. Furthermore, as $\mathcal{G}$ is a Lie subgroup which contains a hyperbolic matrix H then there exists an open neighborhood $\mathcal{U} \subset \mathrm{SL}(2, \mathbb{R})$ of H formed by hyperbolic matrices. Thus, there exist hyperbolic matrices arbitrarily close to the identity map in the image of the map $\mathcal{G} \cap \mathcal{U} \ni \tilde{H} \mapsto \tilde{H} \cdot H^{-1} \in \mathcal{G}$. This implies that for any pair $(A_1, A_2) \in \mathcal{G} \times \mathcal{G}$ satisfying $(\star)$ there exists $\tilde{H}$ arbitrarily close to H so that the matrices $\tilde{H} \cdot H^{-1} \cdot A_1$ and A_2 belong to $\mathcal{G}$ but their eigenvalues do not have the same absolute value. In particular, $\mathcal{A}_{3,\text{loc}}(\mathcal{G})$ is a meager subset of $C^0_{loc}(\Sigma_\kappa, \mathcal{G})$, which proves item 2 (d).

Now we proceed to show that $\mathcal{A}_{2,\text{loc}}(\mathcal{G}) = \mathcal{A}_{3,\text{loc}}(\mathcal{G})$. The inclusion $(\subseteq)$ is immediate. For the converse, observe that if $A \in \mathcal{A}_{3,\text{loc}}(\mathcal{G})$ then for each $n \geqslant 1$ fixed, all matrices $\{A^n(x) : \sigma^n x = x\}$ have the same eigenvalues in absolute value. In particular, either

(i) $|\lambda_i^+| = 1$ for every $1 \leqslant i \leqslant \kappa$, or
(ii) $|\lambda_i^+| = \lambda > \lambda^{-1} = |\lambda_i^-|$ for every $1 \leqslant i \leqslant \kappa$,

where $\lambda_i^+ = \lambda^+(A, \delta_{\underline{i}})$ coincides with the largest eigenvalue of the matrix A_i.

In case (i), it follows from [22, Theorem 2.2] that (up to consider a double cover in case of the existence of invariant non-orientable subbundles) the cocycle A is cohomologous to a Hölder continuous cocycle of the following type:

I. $B(x) = \begin{pmatrix} 1 & \alpha(x) \\ 0 & 1 \end{pmatrix}$, in case A preserves exactly one Hölder continuous orientable subbundle;
II. $B(x) = Id$, if A preserves more than one Hölder continuous orientable subbundle;

III. $B(x) = \begin{pmatrix} \cos\alpha(x) & -\sin\alpha(x) \\ \sin\alpha(x) & \cos\alpha(x) \end{pmatrix}$, whenever A does not preserves any Hölder continuous orientable subbundle.

Therefore, choosing repeatedly x as a fixed point for the shift, one concludes that all matrices A_i are of the same type ie. either all A_i are nilpotent, all are the identity or all are rotation matrices. In any case, all points in the shift space Σ_2 are Lyapunov regular because

$$\lim_{n\to\infty} \frac{1}{n} \log \|A^n(x)^{-1}\|^{-1} = \lim_{n\to\infty} \frac{1}{n} \log \|A^n(x)\| = 0, \qquad \forall\, x \in \Sigma_\kappa.$$

This, together with Proposition 3.1 proves that $A \in \mathcal{A}_{2,\mathrm{loc}}(\mathcal{G})$, as desired.

In case (ii), assume first that $A \in \mathcal{A}_{2,\mathrm{loc}}(\mathcal{G}) \cap \mathcal{H}$. As the cocycle A is hyperbolic then there exists an A-invariant splitting $\{x\} \times \mathbb{R}^2 = E^s_x \oplus E^u_x$, depending Hölder continuously with the point x, and there exist constants $C > 0$ and $\zeta \in (0,1)$ so that $\|A^n(x)\mid_{E^s_x}\| \leqslant C\zeta^n$ and $\|(A^n(x))^{-1}\mid_{E^u_x}\| \leqslant C\zeta^n$ for every $x \in \Sigma_2$ and $n \geqslant 0$. Together with Livšic's theorem (Theorem 1.1), the assumption that all top Lyapunov exponents (resp. smallest Lyapunov exponent) at periodic points are the same and equal to a constant $\log\lambda$ (resp. $-\log\lambda$) implies that there exist Hölder continuous functions $h^s, h^u : \Sigma_2 \to \mathbb{R}$ so that

$$\|A(x)\mid_{E^s_x}\| = \lambda^{-1}\, e^{h^s\circ\sigma(x) - h^s(x)} \qquad \text{and} \qquad \|A(x)\mid_{E^u_x}\| = \lambda\, e^{h^u\circ\sigma(x) - h^u(x)} \tag{4.1}$$

for every $x \in \Sigma_2$. Therefore, if $Q(x)$ is the matrix corresponding to the change of coordinates from the basis determined by the hyperbolic splitting at $E^s_x \oplus E^u_x$ to the basis $\{(e^{-h^u(x)}, 0), (0, e^{-h^s(x)})\}$ then

$$A(x) = Q(\sigma(x))^{-1} \begin{pmatrix} \lambda & 0 \\ 0 & \lambda^{-1} \end{pmatrix} Q(x)$$

for every $x \in \Sigma_2$. The latter shows that $\mathcal{A}_{1,\mathrm{loc}}(\mathcal{G}) \cap \mathcal{H} = \mathcal{A}_{2,\mathrm{loc}}(\mathcal{G}) \cap \mathcal{H} = \mathcal{A}_{3,\mathrm{loc}}(\mathcal{G}) \cap \mathcal{H}$ and proves item 2(b).

In case (ii) it remains to consider the case that all products of the generating matrices A_i $(i = 1, 2)$ are hyperbolic but the cocycle $A \in \mathcal{A}_{2,\mathrm{loc}}(\mathcal{G})$ is not hyperbolic. As the cocycle is not elliptic either (as all eigenvalues of matrices in the semigroup do not intersect the unit circle) and $\kappa = 2$ one has that $A \in \partial\mathcal{E} = \partial\mathcal{H}$. Using Theorem 3.1 (see also Proposition 4 in [29]) we have a dichotomy: either there exists a periodic point $x \in \Sigma_2$ of period k for σ such that $A^k(x)$ is parabolic, or either $E^s_{A_1} = E^u_{A_2}$ or $E^u_{A_1} = E^s_{A_2}$. The existence of a parabolic periodic point is incompatible with the existence of the non-zero Lyapunov exponents $\pm\log\lambda$ at periodic points. If $E^s_{A_1} = E^u_{A_2}$ or $E^u_{A_1} = E^s_{A_2}$, as the eigenvalues of A_1 and A_2 are the same in absolute value, then the eigenvalues of the matrix $A_1 A_2$ lie in the unit circle, a fact which is also incompatible with our assumption. This shows that there exists no such cocycles.

Finally, the previous arguments allow one to reduce the proof of item 2(a) to the characterization of the set $\mathcal{A}_{1,\mathrm{loc}}(\mathcal{G})$ stated at item 2(c). Indeed, given a cocycle $A \in \mathcal{A}_{1,\mathrm{loc}}(\mathcal{G}) \cap \mathcal{H}^c$, all matrices in the semigroup $\mathcal{S}$ are either all rotations or are either all parabolic, and an argument identical to the one used before shows that $\mathcal{A}_{1,\mathrm{loc}}(\mathcal{G}) \cap \mathcal{H}^c$ is a meager subset of $\mathcal{A}_{2,\mathrm{loc}}(\mathcal{G})$. This completes the proof of the theorem. □

Remark 4.1 The assumption that $\kappa = 2$ in Theorem 2.1 was used in one instance in the proof, in order to characterize the boundary of the space of elliptic cocycles. Under the assumption that $\kappa = 2$ it follows from Theorem 3.1 that $\partial\mathcal{E} = \partial\mathcal{H}$. In general it is known that $\partial\mathcal{E}$ may differ from $\partial\mathcal{H}$ (cf. [9] for examples and a more detailed discussion on this problem).

4.2 *Proof of Corollary 2.1*

Consider the potential $\varphi : \Sigma_\kappa \to \mathbb{R}$ given by $\varphi(x) = \log \|A(x)\ |_{F_x}\ \|$, for every $x \in \Sigma_\kappa$. As φ is continuous and the full shift satisfies the specification property, all points in Σ_κ are Lyapunov regular for the cocycle A along the A-invariant subbundle E if and only if there exists $c > 0$ so that the limit

$$\lambda_-(A, x) = \lim_{n\to\infty} \frac{1}{n} \log \|A^n(x)\ |_{F_x}\ \| = \lim_{n\to\infty} \frac{1}{n} \sum_{j=0}^{n-1} \varphi(\sigma^j x)$$

exists everywhere and coincides with $\log c$, for some $c > 1$ (see e.g. [17]). As the cocycle A takes values in $SL(3, \mathbb{R})$ then $\lim_{n\to\infty} \frac{1}{n} \log |\det A(x)\ |_{E_x}\ | = -\log c$ for every $x \in \Sigma_\kappa$.

Therefore, using the observation made at Sect. 3.3, all points in the shift space are Lyapunov regular for the continuous cocycle $A\ |_E$ if and only if all points are Lyapunov regular for the cocycle $B \in C^0_{\mathrm{loc}}(\Sigma_\kappa, SL(2, \mathbb{R}))$ given by

$$B(x) = \frac{1}{|\det A(x)\ |_{E_x}\ |} A(x)\ |_{E_x} \qquad \forall x \in \Sigma_\kappa.$$

The proof of Theorem 2.1 guarantees that either B is hyperbolic (and conjugated to a constant matrix cocycle) or $\lim_{n\to\infty} \frac{1}{n} \log \|B^n(x)v\| = 0$ for every $x \in \Sigma_\kappa$ and $v \in \mathbb{R}^2 \setminus \{0\}$, which is enough to conclude the corollary. qed

Acknowledgments The author is grateful to the organizers of the Workshop New Trends on Lyapunov Exponents - Lisbon 2022, where this work was initiated. The author is deeply grateful to A. Blumenthal, J. Bochi, P. Duarte, F. Micena and J. Siqueira for their questions, comments and references, and to the referee for the careful reading of the manuscript. This work was partially supported by the project 'New trends in Lyapunov exponents' (PTDC/MAT-PUR/29126/2017), by CMUP (UID/MAT/00144/ 2013), and by Fundação para a Ciência e Tecnologia (FCT) - Portugal, through the grant CEECIND/03721/2017 of the Stimulus of Scientific Employment, Individual Support 2017 Call.

References

1. A. Ávila, J. Bochi and J.-C. Yoccoz, Uniformly hyperbolic finite-valued SL(2, $\mathbb{R}$)-cocycles, *Comment. Math. Helv.* 85 (2010), 813–884.
2. L. Backes, Rigidity of fiber bunched cocycles *Bull. Braz. Math. Soc.*, 46:2 (2015) 163–179.
3. I. Bárány, A. Beardon and T. Carne, Barycentric subdivision of triangles and semigroups of Möbius maps. Mathematika 43:1 (1996) 165–171.
4. M. Berger and Y. Wang, *Bounded semi-groups of matrices*, Lin. Alg. Appl. 166 (1992) 21–27.
5. J. Bochi and I. Morris, Continuity properties of the lower spectral radius, *Proc. London Math. Soc.* 110:2 (2015) 477–509.
6. R. Bowen. *Equilibrium states and the ergodic theory of Anosov diffeomorphisms*, volume 470 of *Lect. Notes in Math.* Springer Verlag, 1975.
7. M. Carvalho and P. Varandas, Genericity of historic behavior for maps and flows, *Nonlinearity* 34:10 (2021) 7030–7044.
8. M. Carvalho, V. Coelho, L. Salgado and P. Varandas, Sensitivity and historic behavior on Baire metric spaces, *Ergod. Th Dynam. Sys.* (to appear), https://doi.org/10.1017/etds.2023.3
9. A. Christodoulou, Parameter spaces of locally constant cocycles, *International Mathematics Research Notices*, rnab116, https://doi.org/10.1093/imrn/rnab116
10. R. de la Llave and F. Micena, Lyapunov exponents everywhere and rigidity Journal of Dynamical and Control Systems, 27 (2021) 819–831.
11. P. Duarte and S. Klein, Lyapunov Exponents of Linear Cocycles - Continuity via Large Deviations Atlantis Studies in Dynamical Systems, Series volume 3, Atlantis Press, 2016, 263 pages.
12. G. Ferreira and P. Varandas, Lyapunov "non-typical" behavior for linear cocycles through the lens of semigroup actions, Preprint arXiv:2106.15676
13. H. Furstenberg and H. Kesten. Products of random matrices. *Ann. Math. Statist.*, 31 (1960) 457–469.
14. C. Heil, G. Strang. Continuity of the joint spectral radius: application to wavelets. Linear algebra for signal processing (Minneapolis, MN, 1992), IMA Vol. Math. Appl. 69, p. 51–61, Springer, New York, 1995.
15. K. Iwasawa, *On some types of topological groups*, Ann. of Math., 50 (1949) 507–558.
16. B. Kalinin. Livšic theorem for matrix cocycles. *Ann. of Math.* (2) 173 (2011), 1025–1042.
17. J. Li and M. Wu, Generic property of irregular sets in systems satisfying the specification property. *Discrete Cont. Dynam Sys.*, 34: 2 (2014) 635–645.
18. A. N. Livšic, Homology properties of Y-systems. *Math. Zametki* 10 (1971) 758–763.
19. A. N. Livšic. Cohomology of dynamical systems. *Math. USSR Izvestija* 6 (1972), 1278–1301.
20. V. Oseledets, Multiplicative ergodic theorem. Lyapunov characteristic numbers for dynamical systems. *Trans. Moscow Math. Soc.* 19 (1968) 197–221.
21. W. Parry and M. Pollicott. The Livšic cocycle equation for compact Lie group extensions of hyperbolic systems. *J. Lond. Math. Soc.* (2) 56(2) (1997), 405–416.
22. V. Sadovskaya, Cohomology of $GL(2, \mathbb{R})$-valued cocycles over hyperbolic systems. *Discrete Contin. Dyn. Sys.* 33 (2013), no. 5, 2085–2104.
23. V. Sadovskaya, Cohomology of fiber bunched cocycles over hyperbolic systems. *Ergod. Th. & Dynam. Sys.* 35 (2015), no. 8, 2669–2688.
24. K. Schmidt. Remarks on Livšic theory for non-abelian cocycles. *Ergod. Th. & Dynam. Sys.* 19(3) (1999), 703–721.
25. J. Suárez and T. Moreno, The limit property for the interior solid angles of some refinement schemes for simplicial meshes. *J. Comput. Appl. Math.* 275 (2015) 135–138.

26. X. Tian, Nonexistence of Lyapunov exponents for matrix cocycles. *Ann. Inst. Henri Poincaré - Probab. Stat.* 53 (2017), no. 1, 493–502.
27. P. Varandas, On the cohomological class of fiber-bunched cocycles on semi simple Lie groups, *Rev. Mat. Iberoam.* 36:4 (2020) 1113–1132.
28. M. Viana, Lectures on Lyapunov Exponents, Cambridge University Press, Rio de Janeiro, 2014.
29. J.-C. Yoccoz, Some questions and remarks about SL(2, $\mathbb{R}$)-cocycles. In: Brin, M., Hasselblatt, B., Pesin, Y. (eds.) *Modern Dynamical Systems and Applications*, pp. 447–458. Cambridge University Press, Cambridge, 2004.